Magnetic Materials and Magnetic Levitation

Edited by Dipti Ranjan Sahu
and Vasilios N. Stavrou

Published in London, United Kingdom

IntechOpen

Supporting open minds since 2005

Magnetic Materials and Magnetic Levitation
http://dx.doi.org/10.5772/intechopen.80191
Edited by Dipti Ranjan Sahu and Vasilios N. Stavrou

Contributors
Peng Li, Laith Alahmed, Jiahao Liu, Liang Fang, Kuldeep Chand Verma, Yared Worku, Diptiranjan Sahu, V. V. Srinivasu, Ahu Arslan Yildiz, Vadim Glebovsky, Poorva Sharma, Ashwini Kumar, Guolong Tan, Jingyou Tang, Kei Hayashi, Hezhang Li, Yoshimi Nagashima, Yuzuru Miyazaki, Mao Eguchi, Anca Birsan, Victor Kuncser, Fatih Ozefe

First published in London, United Kingdom, 2021 by IntechOpen
IntechOpen is the global imprint of INTECHOPEN LIMITED, registered in England and Wales, registration number: 11086078, 5 Princes Gate Court, London, SW7 2QJ, United Kingdom
Printed in Croatia

British Library Cataloguing-in-Publication Data
A catalogue record for this book is available from the British Library

Additional hard and PDF copies can be obtained from orders@intechopen.com

Magnetic Materials and Magnetic Levitation
Edited by Dipti Ranjan Sahu and Vasilios N. Stavrou
p. cm.
Print ISBN 978-1-83962-160-4
Online ISBN 978-1-83962-161-1
eBook (PDF) ISBN 978-1-83962-162-8

For EU product safety concerns:
IN TECH d.o.o., Prolaz Marije Krucifikse Kozulić 3, 51000 Rijeka, Croatia, info@intechopen.com or visit our website at intechopen.com.

Meet the editors

Dr. Dipti Ranjan Sahu is currently an Associate Professor of Physics in the Department of Natural and Applied Sciences at the Namibia University of Science and Technology (NUST). He received his doctoral degree in physics from the Institute of Materials Science, Utkal University, India. After his Ph.D., he worked as a postdoctoral researcher and a visiting scientist in several institutions such as the National Taiwan University, National Cheng Kung University (Taiwan), and the University of Witwatersrand (South Africa). His research focus is on multifunctional materials including nano-materials, ceramic, composites, spintronics, magnetic materials, and the application of functional materials in devices. He has published more than 90 international peer-reviewed journal articles, 1 book, 2 book chapters, and more than 100 research articles in conference proceedings and meetings.

Dr. V. N. Stavrou is currently an Adjunct Member at the Hellenic Naval Academy, Piraeus, Greece. He received an MSc and a Ph.D. in theoretical solid-state physics from the University of Essex in England, in 1995 and in 1999 respectively. He has held postdoctoral positions at the following research institutions: a) Deutsche Forschungsanstalt fuer Luft und Raumfahrt e.V (German Aerospace Research Center) in Germany, b) Helsinki University of Technology, c) State University of New York (SUNY) at Buffalo, USA and d) University of Iowa, USA. He is specifically interested in searching the electronic, optical, and lattice properties of semiconducting low dimensional structures.

Contents

Preface

Magnetic materials have acquired a very important position in several high-tech areas and technological development. Such materials are being classified not only based on their origin but also by the nature of processing, properties, functions, and applications. *Magnetic Materials and Magnetic Levitation* present the basics of magnetism, magnetic materials, magnetic structures, and their applications in device technologies. New magnetic materials and hybrid structures are developed using different synthesis and fabrication techniques. Different phenomena and interesting properties are studied theoretically and experimentally using advanced characterization techniques. Magnetic materials are now the building block of all technological innovation. This book presents an overview of different magnetic materials including theoretical study, synthesis, characterization, and application of magnetic materials. The chapters and different topics of the book will provide a key understanding of different magnetic materials. It will be very helpful to students, researchers, academicians, and professionals. The book is related to magnetic materials and gives an idea and enthusiasm to readers for scientific advances in the subject and technology.

Dipti Ranjan Sahu
Namibia University of Science and Technology,
Namibia

Vasilios N. Stavrou
Hellenic Naval Academy,
Greece

Section 1

Magnetic Materials

Chapter 1

Structural, Electrical, and Magnetic Properties of Mullite-Type Ceramic: $Bi_2Fe_4O_9$

Poorva Sharma, Ashwini Kumar, Jingyou Tang and Guolong Tan

Abstract

This work reports the structure, electrical and magnetic properties of the orthorhombic (*Pbam*) structured bulk $Bi_2Fe_4O_9$ synthesized by the solid-state reaction process. $Bi_2Fe_4O_9$ has been studied using several experimental techniques such as X-ray diffraction, scanning electron microscopy, Raman spectroscopy, dielectric spectroscopy, and magnetometry. Rietveld-refined X-ray diffraction data and Raman spectroscopy results clearly reveal the formation of $Bi_2Fe_4O_9$ perovskite structure and all the peaks of $Bi_2Fe_4O_9$ perfectly indexed in the orthorhombic (*Pbam*) structure. It has been established that the Raman spectrum identified A_g, B_{2g}, and B_{3g} active optical phonon modes, and that the Raman peak at 470 cm^{-1} may have a magnetic origin. As a result, the coexistence of weak ferromagnetic and antiferromagnetic orders in $Bi_2Fe_4O_9$ ceramic was established. The remanent magnetization ($2M_r$) and coercivity ($2H_c$) are 8.74×10^{-4} emu/g and 478.8 Oe, respectively. We report a remarkable multiferroic effects in polycrystalline $Bi_2Fe_4O_9$ ceramic. These characteristics make this material very useful in technology and practical applications.

Keywords: multiferroic, $Bi_2Fe_4O_9$, Raman spectroscopy, ferroelectric, magnetic measurement

1. Introduction

Multiferroic materials exhibit more than one primary ferroic order parameters (i.e. ferroelectricity, ferroelasticity and ferromagnetism) in same phase which was first proposed by Schmid in 1994 [1]. In recent years, there has been a strong interest in systems that exhibit convergence between magnetic degrees of freedom, electronic degrees of freedom, and orbital degrees of freedom. Perovskite based oxides have attracted much attention due to their interesting structural, magnetic, optical and electronic properties [2]. A large number of publications have been devoted to multiferroic materials working with theory, experimentation and application features. Bismuth-based complex oxides ($Bi_2Fe_4O_9$) with mullite-type structure, as an important active material, has a wide application prospect in the fields of magnetic recording media, sensor, magnetoresistive devices, solid oxide fuel cell, scintillators and photocatalyst [3–7].

The crystallographic structure of $Bi_2Fe_4O_9$ is orthorhombic with space group *Pbam*, No. 55, which belongs to the mullite-type crystal structure family [8, 9].

A unit cell of $Bi_2Fe_4O_9$ consists of two formula units with an equal distribution of Fe ions between the edge-sharing octahedral (FeO_6) and corner-sharing tetrahedral (FeO_4) positions with Bi^{3+} ions are surrounded by eight oxygen atoms. Bulk $Bi_2Fe_4O_9$ synthesized by solid state reaction exhibiting an antiferromagnetic (AFM) ordering at T_N = 260 K and ferroelectric (FE) hysteresis loops at T = 250 K, which indicates that $Bi_2Fe_4O_9$ is a promising multiferroic material [9, 10]. An unexpected multiferroic effect, which was observed as a coexistence of AFM and FE polarization, was reported in $Bi_2Fe_4O_9$, attributed to frustrated spin system coupled with phonons [10]. Low electrical conductivity in ferrites is useful for inductor, transformer cores and in switch mode power supplies. On the other hand, studies of electric and dielectric properties are also equally important from both fundamental and application point of view. Dielectric and magnetic behavior of ferrites is greatly influenced by an order of magnitude of conductivity and is mostly dependent on preparation method and sintering conditions [11].

Although, due to search of new multifunctional materials, the recent work carried out is the very important and needed [12–15]. Rao et al. reported the multifunctional properties of mullite-type structured Nd-doped $Bi_2Fe_4O_9$ and the spin-orbital coupling by D-M interactions enhances the ferromagnetic (FM) behavior of the Nd-doping $Bi_2Fe_4O_9$ [12]. Ameer et al. studied the structural, electronic, and magnetic properties of $Bi_2Fe_4O_9$ with different magnetic ordering using the projector augmented wave (PAW) method based on density functional theory (DFT). They proposed that the FM $Bi_2Fe_4O_9$ is a semiconductor with an indirect optical bandgap of 1.732 eV and the exchange mechanism started to work, resulting in the exchange splitting in $Bi_2Fe_4O_9$, while the antiferromagnetic (AFM) $Bi_2Fe_4O_9$ is a multiband semiconductor without splitting of the majority and minority spin states [13]. In another study, the researchers believed that Zn substitution in $Bi_2Fe_4O_9$ would induce p-type conductivity, suggesting that $3d$ transition metal ions doping in $Bi_2Fe_4O_9$ provides the capabilities to develop low-bandgap, heterojunction-based optoelectronic devices [14]. In addition, Pooladi et al. studied the $Bi_2Fe_{4-x}Mn_xO_9$ ($0.0 \leq x \leq 1.0$) nanoparticles synthesized by reverse chemical co-precipitation method. With increase in Mn concentration, the coercivity of the nanoparticles enhances significantly and the saturation magnetization decreased. Also, the Mn substitution at Fe site in $Bi_2Fe_{4-x}Mn_xO_9$ increases the dielectric constant [15]. These types of structures and materials are interested due to their structural, magnetic properties and the relationship between orbital, spin and charge degrees of freedom.

Various chemical methods such as solid-state reaction route, chemical co-precipitation, sol-gel and hydrothermal have been used to produce $Bi_2Fe_4O_9$ [14–16]. The properties of materials are highly dependent on structural, microstructural properties and methods of synthesis. In this regard, it is of interest to develop controlled methods for making materials in oxide forms for further functional applications. Thermal heating in the oxygen atmosphere at high temperatures contributes to the oxidation process and formation of oxide forms, which has a significant impact on physical, chemical and magnetic properties of compounds [17–21]. Zdorovets et al. reported the systematic study of the effect of thermal annealing on changes in the structural properties and phase compositions of metallic cobalt based nanostructures [17]. Rusakov et al. described the effect of thermal annealing on structural and magnetic characteristics, as well as phase transformations in Fe–Ni/Fe–Ni–O nanoparticles. They found that the initial nanoparticles were a three-phase system consisting of Fe–Ni–O oxide with spinel structure and a Fe–Ni alloy with face-centered and body-centered cubic lattices. As a result of thermal annealing, the decrease in the Fe–Ni phase is associated with the subsequent ordering of the Fe–Ni–O phase with a decrease in the crystal lattice parameter and an increase in the degree of crystallinity [19]. If annealing is carried out in air, the phase transition related to the structural transformation of iron oxide is in the range of 600–1000°C due to the change of thermal vibration of atoms in the lattice node, the annealing of point defects and the introduction of oxygen at high temperature [20].

Although $Bi_2Fe_4O_9$ has obvious importance as a functional material, there are few reports in the literature. Here, we present the structural and physical properties of bulk $Bi_2Fe_4O_9$ ceramic synthesized by a solid-state reaction route. One needs detailed knowledge of the crystal structure to understand the physical properties. Therefore, we aimed to understand the crystal structure by X-ray powder diffraction followed by Rietveld refinement using FullPROF program [22]. In addition, $Bi_2Fe_4O_9$ was subsequently characterized using several experimental techniques, such as Raman spectroscopy, SEM, dielectric and ferroelectric spectroscopy, and magnetometry, which are discussed in detail.

2. Experimental details

Bulk $Bi_2Fe_4O_9$ ceramic was synthesized through solid-state reaction route (SSR). The SSR is a commonly used synthesis method for obtaining polycrystalline bulk materials from solid reagents. This method provides a great deal of choices for starting materials like oxides, carbonates, etc. Since solids do not react with each other at room temperature, very high temperatures are usually employed to allow appropriate reaction to occur at a significant rate. Therefore, both thermodynamic and kinetic factors are important in SSR. In the SSR method, the solid reactants undergo a chemical reaction at high temperature in the absence of any solvent, thereby producing a stable product. High purity Bi_2O_3, Fe_2O_3 were carefully weighed and stoichiometrically mixed in an agate mortar for 5 hours. The powder was doubly thermally calcined consecutively at 650°C for 1 hour and 850°C for 6 hours with intermediate grinding in oxygen-containing medium. Finally, pellets were sintered at 850°C for 6 hours, resulting in good densification. Thermal heating (i.e. calcination and annealing) is a mean of controlling the structural changes, properties, and phase compositions [23]. In this case, introduction of oxygen leads to the formation of oxide compounds. For crystallinity and phase identification X-ray diffraction (XRD) pattern were taken using CuKα1 radiation (λ = 1.5406 Å) of a Bruker D8 Advance X-ray diffractometer. Crystal structure characterization of synthesized sample was performed by employing Rietveld whole profile fitting method using FullPROF software [22].

The sample quality, morphology, grain distribution, density/voids in the samples were studied with scanning electron microscope (JEOL, JSM-5600). Raman measurements on as synthesized sample was carried out on Jobin-Yovn Horiba LABRAM (System HR800) spectrometer with a 632.8 nm excitation source equipped with a Peltier cooled CCD detector. Dielectric measurements were made as a function of frequency in the range of 100 Hz–1 MHz on Novocontrol alpha-ANB impedance analyzer at room temperature. Ferroelectric measurement was carried out using a ferroelectric loop tracer based on Sawyer-Tower circuit. The *M-H* curve was performed using a Lakeshore VSM 7410 model.

3. Results and discussion

3.1 Crystal structure analysis

The room temperature XRD pattern of bulk $Bi_2Fe_4O_9$ sample is shown in **Figure 1(a)**. From the XRD pattern we can index the data in orthorhombic phase as shown in **Figure 1(a)**. The present XRD patterns matches with JCPDS

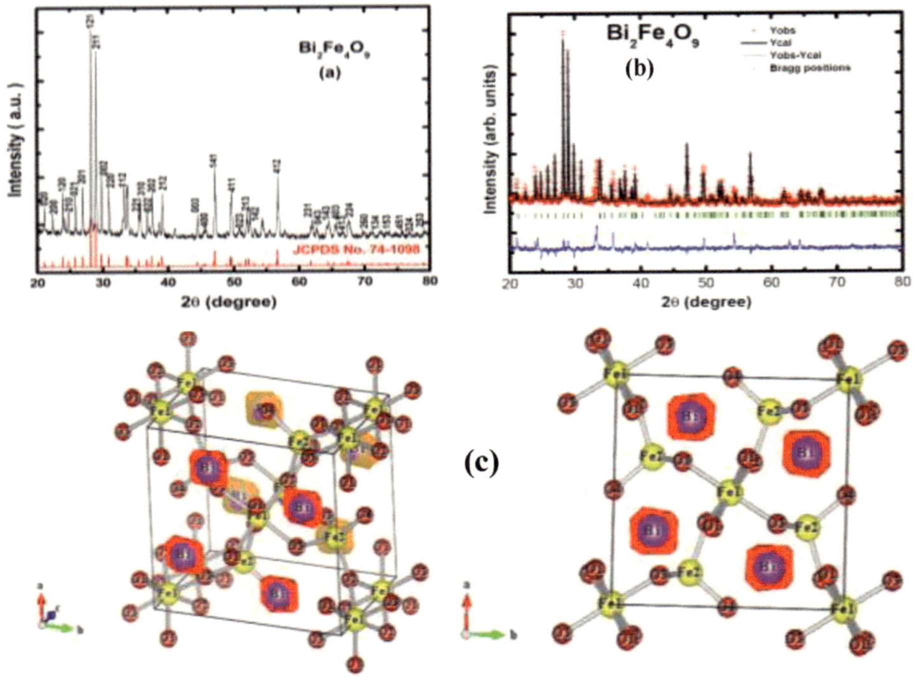

Figure 1.
(a) Measured powder X-ray diffraction pattern for $Bi_2Fe_4O_9$ ceramic at room temperature, (b) Rietveld-refined XRD pattern for $Bi_2Fe_4O_9$, and (c) symmetric representation of the unit cell for $Bi_2Fe_4O_9$ ceramic viewed in abc *and* ab *planes.*

#01-74-1098 ($Bi_2Fe_4O_9$) [24]. In order to further confirm structural data, Rietveld refinement of the XRD pattern for $Bi_2Fe_4O_9$ sample was performed using FullPROF program and shown in **Figure 1(b)**. The composition of phase and its concentration in the structure have been determined using the Rietveld method, which is based on the estimation of the diffraction peak area and the analysis of their contributions to the entire X-ray diffraction. It should be noted here that XRD pattern having a small secondary phase peaks corresponding to the Fe_2O_3 and its phase concentration is less than 2%, which does not affect the measured properties of studied ceramic. The XRD pattern of parent $Bi_2Fe_4O_9$ was refined with orthorhombic (*Pbam*) structure with lattice parameters a = 7.941(4) Å, b = 8.420(4) Å and c = 5.986(4) Å. The obtained lattice parameters are consistent with earlier reported data [25]. The Rietveld-refined calculated parameters of $Bi_2Fe_4O_9$ are documented in **Table 1**. We have illustrated structural parameters for $Bi_2Fe_4O_9$ ceramic, and also identify the residuals for weighted pattern R_{wp}, the expected weighted profile factor R_{exp}, and goodness of fit χ^2. The selected bond lengths and bond angles are mentioned in **Table 1**. The average value of the Bi–O bond is 2.482 Å. The generated orthorhombic structure of $Bi_2Fe_4O_9$ ceramic is depicted in **Figure 1(c)**. In the crystal structure, chains of FeO_6 octahedra parallel to the c axis are connected *via* FeO_4 tetrahedra alternating with bismuth atoms along the c axis.

The symmetric pseudo-Voigt functions are used to calculate the degree of crystallinity based on the estimation of the diffraction width and shapes. We have measured the full width half maxima (FWHM) of the recorded diffraction lines, which allowed us to characterize the perfection of the crystal structure and evaluate the degree of crystallinity [21]. The value of % of crystallinity for $Bi_2Fe_4O_9$ = 81.3% was calculated using the formula:

$$Crystallinity = \frac{Area\ of\ Crystalline\ peaks}{Area\ of\ all\ peaks\left(i.e\ Crystalline + Amorphous\right)} \times 100 \quad (1)$$

The *distortion of the crystal lattice* (ρ_{dil}), which characterizes the number of defective or porous inclusions in the structure as a result of external influences, was found according to formula:

$$\rho_{dil} = \left(1 - \frac{\rho}{\rho_o}\right) \times 100\% \quad (2)$$

Here, ρ_0 is the density of the reference sample taken from the JCPDS #01-74-1098 database, ρ is the calculated density of the sample. Calculated value of distortion of the crystal lattice is ρ_{dil} = 3.1%.

We used the Williamson-Hall plot to observed the effect of the phase composition on distortions and deformation of the crystal structure of $Bi_2Fe_4O_9$ ceramic, based on estimating the angular dependence of the full width at half maximum (FWHM) of diffraction lines (**Figure 2**). We obtained the strain (ε) value for $Bi_2Fe_4O_9$ ceramic is 0.00412 ± 0.0016. Thermal heating at high temperature helps to reduce the distortion value [19].

a = 7.941(4) Å		b = 8.420(4) Å	c = 5.986(4) Å	V = 400.31(2) Å^3
Atoms	x	y	z	***R*-values**
Bi	0.3230	0.1745	0.0000	R_{wp} = 7.36%
Fe_1	0.0000	0.0000	0.2582	R_{exp} = 3.41%
Fe_2	0.1465	0.3360	0.5000	R_p = 4.61%
O_1	0.3485	0.4292	0.0000	χ^2 = 4.65%
O_2	0.3671	0.4047	0.5000	*GoF* = 2.2
O_3	0.1312	0.2054	0.2413	
O_4	0.0000	0.5000	0.5000	
Bond type	**Bond length (Å)**		**Bond type**	**Bond angle (°)**
Bi–O_1	2.153		O_1–Bi–O_1	151.93
Bi–O_3	3.017		O_3–Bi–O_3	86.06
Fe_1–O_1	2.047		Bi–O_1–Bi	141.13
Fe_1–O_2	1.962		O_2–Fe_2–O_3	103.98
Fe_1–O_3	2.022		O_3–Fe_2–O_3	108.93
Fe_2–O_1	3.485		O_3–Fe_2–O_4	113.59
Fe_2–O_2	1.846		Fe_1–O_1–Fe_1	98.02
Fe_2–O_3	1.901		Fe_1–O_2–Fe_1	94.99
Fe_2–O_4	1.805		Fe_1–O_2–Fe_2	129.72
			Fe_1–O_3–Fe_2	119.09
			Fe_2–O_4–Fe_2	172.00

Table 1.
Rietveld-refined room temperature structural parameters, important bond lengths, and bond angles for $Bi_2Fe_4O_9$ ceramic from XRD.

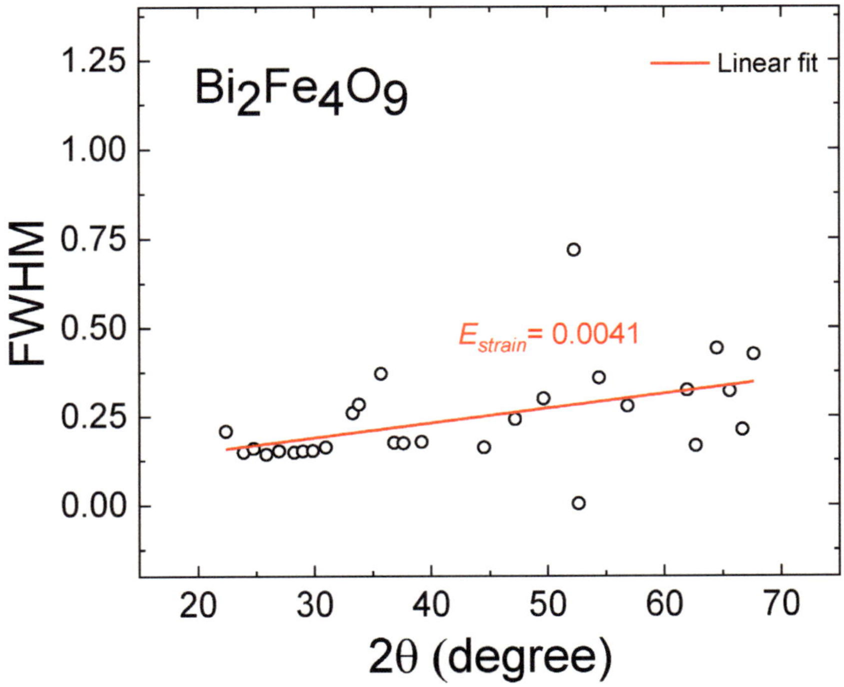

Figure 2.
Williamson-Hall plot for $Bi_2Fe_4O_9$ ceramic.

3.2 SEM analysis

The surface morphological and microstructural properties of $Bi_2Fe_4O_9$ compound was investigated using scanning electron microscopy (SEM). **Figure 3** (upper part) shows the SEM micrograph of $Bi_2Fe_4O_9$ thermally sintered at 850°C for 10 hours. Typical SEM image shows that microstructures comprising of non-uniform distribution of grains with an estimated average grain size of 1.5 μm indicating polycrystalline nature. Even though the SEM image shows that there are some pores between loosely connected grains in the sample. The surface area of a catalyst is a key aspect to determine the adsorption capacity of reactants on the catalyst surface [26]. We have measured the active surface area using a Brunauer-Emmett-Teller (BET) measurement system at 77 K through nitrogen adsorption-desorption isotherm method. The BET active surface area of $Bi_2Fe_4O_9$ is 1.2 m^2/g, which is in good agreement with the values reported in the literature [27]. In order to obtain photocatalytic efficiency, it is necessary to increase the specific surface area by doping or reducing grain sizes. In addition, we have measured the material's apparent density which is defined as the mass per unit volume of the material in absolute dense condition [28]. The obtained density of the present calcined $Bi_2Fe_4O_9$ ceramic is 6.51 g/cm^3 which match well with the density for the $Bi_2Fe_4O_9$ (ρ = 6.48 g/cm^3) from reference file: JCPDS card number 74-1098.

3.3 Raman scattering analysis

Raman scattering spectroscopy has been extensively utilized to study the crystal lattice vibrations. Raman scattering spectroscopy would also offer a distinctive potential as a sensitive probe for the spin dynamics and studying the effect of magnetic ordering. Raman spectrum of $Bi_2Fe_4O_9$ at room temperature is depicted in lower part of **Figure 3**. The Raman active modes of the structure can be

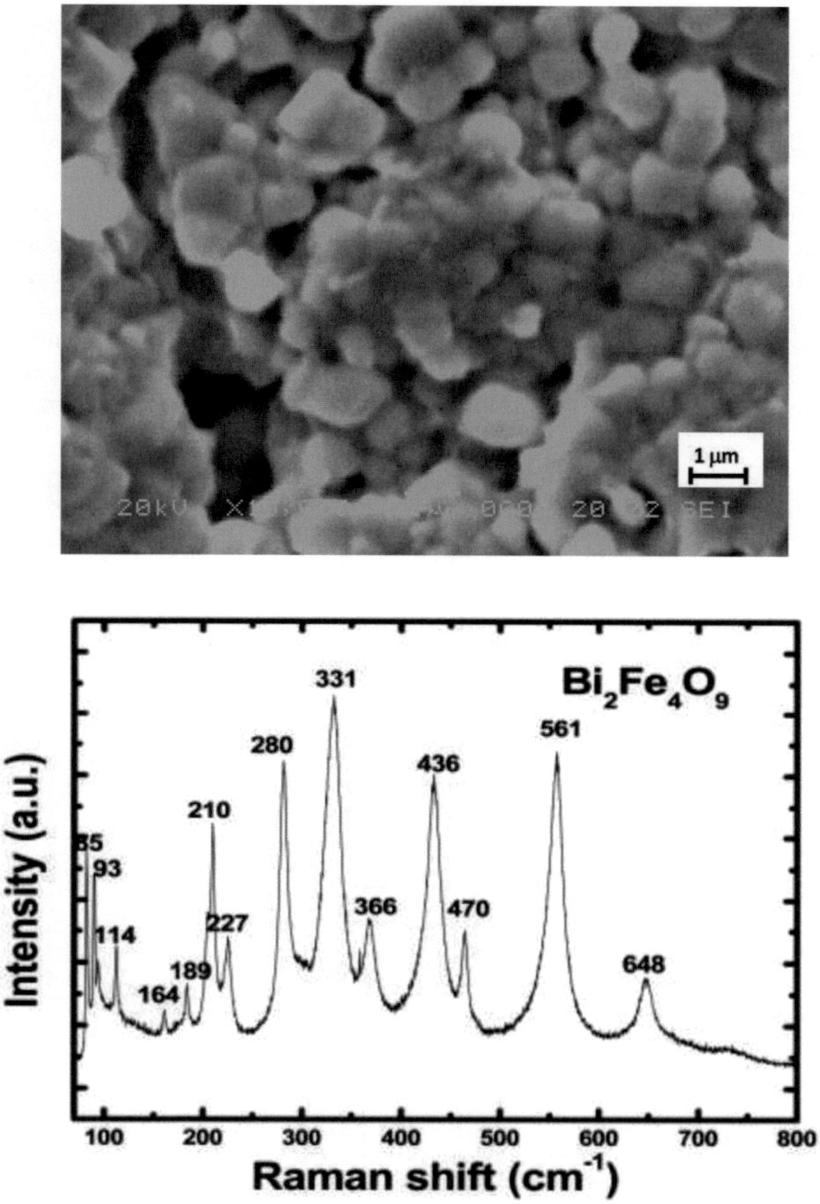

Figure 3.
(Upper) Scanning electron microscope images of $Bi_2Fe_4O_9$ and (lower) Raman spectra for $Bi_2Fe_4O_9$ mullite-type ceramic carried out at room temperature.

summarized using the irreducible representation $12A_g + 12B1_g + 9B2_g + 9B3_g$, which is employed to describe Raman modes of orthorhombic (*Pbam* space group) [9]. In the measured Raman spectra all are the A_g modes (85, 93, 114, 210, 227, 280, 331, 366, 436, 465, 561, and 648 cm^{-1}) accept modes in attendance at 164 (B_{2g}) and 189 (B_{3g}) cm^{-1}. The agreement between experimental and predicted values is relatively good for the all frequency modes, dominated by Bi vibrations. The Raman peak centered at 470 cm^{-1} is might be attributed to magnetic ordering effect on phonon line width consistent with earlier observation of bands at ~260 and 472 cm^{-1} due to magnon scattering [9]. It would be more practical to study the magnetic excitations in $Bi_2Fe_4O_9$ under the assumption that they involve two-magnon processes, like in the well-known cases of ferrites [29] or cuprates [30, 31]. At higher frequency

(>250 cm^{-1}), it is unlikely the magnetic-order-induced bands correspond to one-magnon excitations but in rare-earth orthoferrites ($RFeO_3$; R = Dy, Ho, Er, Sm, etc.) have frequencies below 25 cm^{-1} for comparison the zone-center magnons [32].

3.4 Dielectric and P-E loop studies

The real part of permittivity (ε') and loss tangent ($\tan\delta$) as a function of frequency of $Bi_2Fe_4O_9$ ceramics near at surrounding temperature is shown in **Figure 4(a)** and **(b)**. The value of ε' and $\tan\delta$ for $Bi_2Fe_4O_9$ are about 21.57 and 0.05, respectively at frequency 10 Hz. At higher frequency (~1 MHz) the value of ε' and $\tan\delta$ are 18.59 and 0.006, respectively. Dielectric behavior (i.e. ε' and $\tan\delta$) decreases with increase in frequency and it is constant at higher frequency region. From **Figure 4(a)** and **(b)** we have found that the value of dielectric constant in the

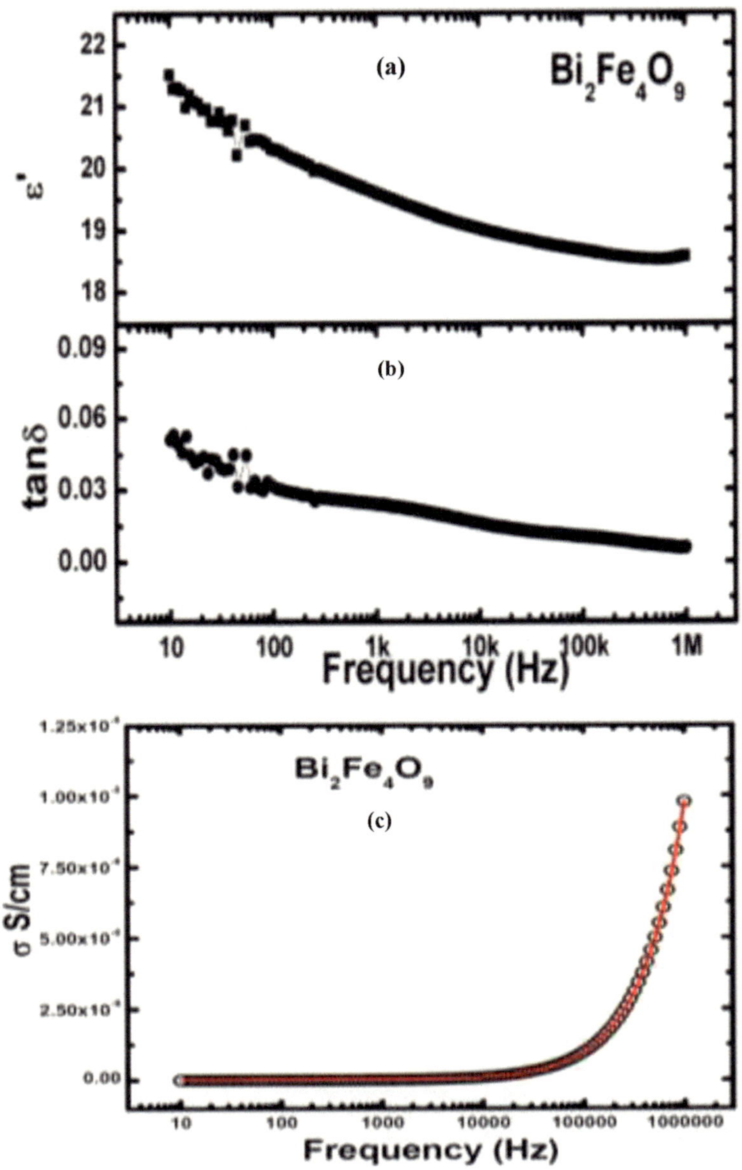

Figure 4.
(a) Dielectric constant, (b) dielectric loss, and (c) frequency-dependent conductivity of $Bi_2Fe_4O_9$ ceramic.

whole frequency range (10 Hz–1 MHz) is nearly constant representing the low loss in the prepared ceramic. This result appears to be consistent with previous empirical analysis using the Maxwell-Wagner model with thermal activation across multiple band gaps in isolated impurities [15, 33]. **Figure 4(c)** shows the semilog plot of conductivity (σ) at room temperature with frequency. The study of the frequency dependence of the conductivity is a deep-rooted method for describing the hopping dynamics of the charge carrier. The conductivity plot exhibits both low and high frequency dispersion phenomena [34–37]. The low-frequency region corresponds

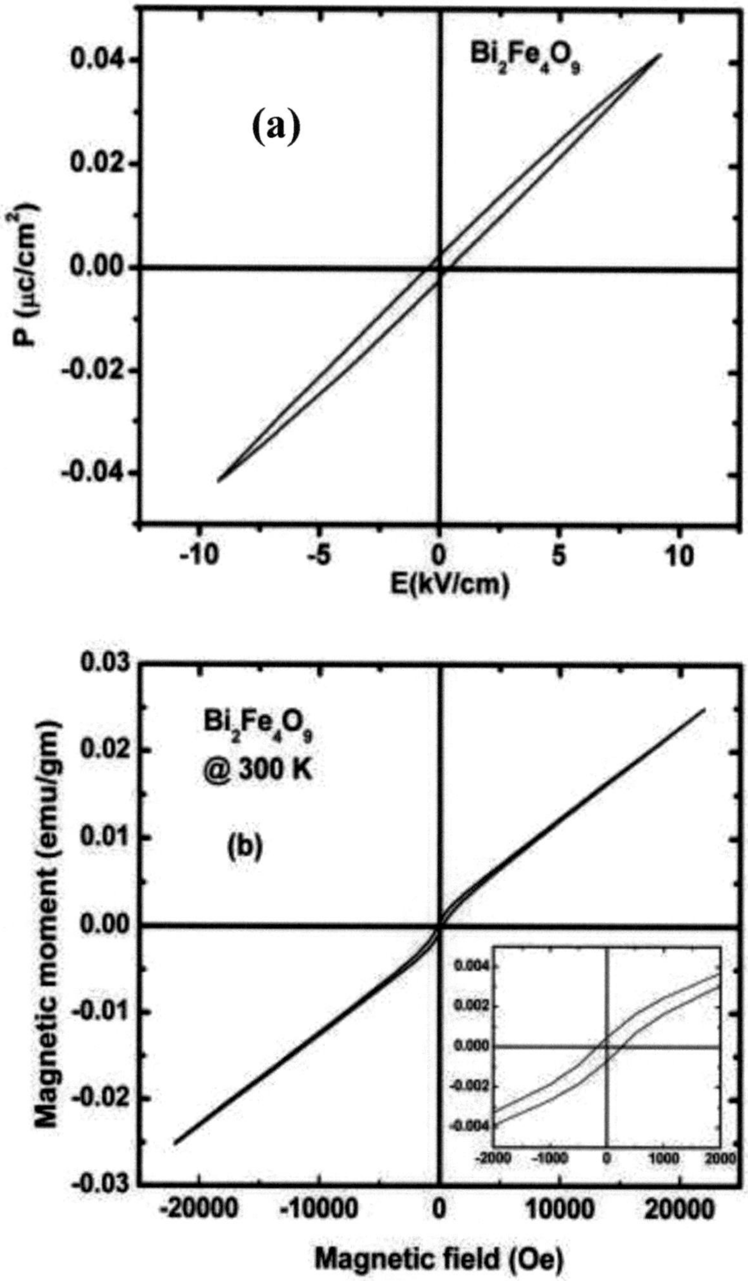

Figure 5.
(a) Polarization hysteresis P-E loop and (b) field-dependent magnetic hysteresis loop of $Bi_2Fe_4O_9$ ceramic at room temperature.

to the *dc* conductivity (σ_{dc}), which is due to the band conduction, and it is frequency independent. The high-frequency region corresponds to the *ac* conductivity (σ_{ac}), which is frequency dependent. To conclude, the electrical conductivity σ for $Bi_2Fe_4O_9$ follows the Jonscher power law [37]: $\sigma_{ac}(\omega) = \sigma_{dc} + A(T)\omega^n$. Here, A is the pre exponential factor and n is the power law exponent. The exponent n can have a value between 0 and 1. This parameter is frequency independent but temperature and material dependent.

Ferroelectric hysteresis $P(E)$ loop of the $Bi_2Fe_4O_9$ ceramic at room temperature represented in **Figure 5(a)**. The obtained loop indicates that there are ferroelectric properties with finite remanent polarization with the applied electric field in the prepared sample. Under an electric field of up to 10 kV/cm, the remanent polarization ($2P_r$) of $Bi_2Fe_4O_9$ was found to be 0.006 μC/cm^2. The observed polarization values are closely consistent with literature results [38].

3.5 Magnetic analysis (*M-H* curve)

From the measured *M-H* loop of $Bi_2Fe_4O_9$ ceramic (**Figure 5(b)**), the magnetic parameters we obtained are remanent magnetization ($M_r = 4.37 \times 10^{-4}$ emu/g), coercivity ($H_c = 239.4$ Oe) and saturation magnetization ($M_s = 0.024$ emu/g). In our $Bi_2Fe_4O_9$ ceramic, antiferromagnetic (AFM) and weak ferromagnetic (WFM) interactions exist simultaneously are consistent with the data reported earlier [15, 38–40]. The WFM order can be seen in the low magnetic field region. WFM order itself can be understood as a result of canted spin arrangements in two sublattices [41]. As the magnetic field increases, the ferromagnetic order saturates and the antiferromagnetic component dominates. There is even no sign of saturation. Obviously, in the prepared ceramic, M_r and M_s have achieved a non-zero value. We may note that the measured hysteresis curve confirm that the relationship between the applied magnetic field and the magnetization does not evidence of a linear behavior and shows the WFM. In the future, we can improve the magnetic and electric properties of $Bi_2Fe_4O_9$ ceramics with the appropriate doping or preparation techniques.

4. Conclusions

We have successfully synthesized polycrystalline $Bi_2Fe_4O_9$ through solid-state reaction route. X-ray diffraction pattern confirmed the formation of $Bi_2Fe_4O_9$ with the orthorhombic structure (space group *Pbam*). The obtained density of the present calcined $Bi_2Fe_4O_9$ ceramic is 6.51 g/cm^3. The room temperature Raman peak at 470 cm^{-1} is may be due to the magnetic origin of the material and might be related to magnon scattering. The value of the dielectric constant seems to be non-variable over the entire frequency range. It has been observed that the sample showed the presence of ferroelectricity. This study conclusively reported that a significant magnetic and ferroelectric order coexist in polycrystalline $Bi_2Fe_4O_9$ at room temperature. There is hysteresis in the electrical and magnetic curves, which proves the multiferroic nature of $Bi_2Fe_4O_9$ and making it more suitable for advanced technology and practical applications.

Acknowledgements

We acknowledge the financial support from National Natural Science Foundation of China under grant numbers 11774276 and 51074131. The authors are grateful to Dr. S. Satapathy for their long-term collaboration and numerous fruitful discussions.

Conflicts of interest

All the authors declare that they have no conflict of interest.

Author details

Poorva Sharma[1*], Ashwini Kumar[1*], Jingyou Tang[1] and Guolong Tan[1,2]

1 Key Laboratory of Multifunctional Materials, Department of Electronic Engineering, Luzhou Vocational and Technical College, Luzhou, Sichuan, China

2 Institute of New Materials, Wuhan University of Technology, Wuhan, China

*Address all correspondence to: poorva@nuaa.edu.cn and 101101216@seu.edu.cn

References

[1] Schmid H. Some symmetry aspects of ferroics and single phase multiferroics. Journal of Physics. Condensed Matter. 2008;**20**:434201

[2] Roni P. The Perovskite Handbook. Israel: Metalgrass Ltd; 2018. p. 108

[3] Park T-J, Papaefthymiou GC, Moodenbaugh AR, Mao Y, Wong SS. Synthesis, and characterization of submicron single-crystalline $Bi_2Fe_4O_9$ cubes. Journal of Materials Chemistry. 2005;**15**:2099-2105

[4] Liu H, Li L, Guo C, Ning J, Zhong Y, Hu Y. Thickness-dependent carrier separation in $Bi_2Fe_4O_9$ nanoplates with enhanced photocatalytic water oxidation. Chemical Engineering Journal. 2020;**385**:123929

[5] Yang H, Dai J, Wang L, Lin Y, Wang F, Kang P. A novel approach to prepare $Bi_2Fe_4O_9$ flower-like spheres with enhanced photocatalytic performance. Scientific Reports. 2017;**7**:768

[6] Poghossian AS, Abovian HV, Avakian PB, Mkrtchian SH, Haroutunian VM. Bismuth ferrites: New materials for semiconductor gas sensors. Sensors and Actuators B: Chemical. 1991;**4**:545

[7] Shamir N, Gurewitz E. The magnetic structure of $Bi_2Fe_4O_9$ – analysis of neutron diffraction measurements. Acta Crystallographica. Section A. 1978;**34**:662

[8] Niizeki N, Wachi M, Fur Z. The crystal structures of $Bi_2Mn_4O_{10}$, $Bi_2Al_4O_9$ and $Bi_2Fe_4O_9$. Kristallografiya. 1968;**127**:173-187

[9] Iliev MN, Litvinchuk AP, Hadjiev VG, Gospodinov MM, Skumryev V, Ressouche E. Phonon and magnon scattering of antiferromagnetic $Bi_2Fe_4O_9$. Physical Review B. 2010;**81**:024302

[10] Singh AK, Kaushik SD, Kumar B, Mishra PK, Venimadhav A, Siruguri V, et al. Substantial magnetoelectric coupling near room temperature in $Bi_2Fe_4O_9$. Applied Physics Letters. 2008;**92**:132910

[11] Verma A, Goel TC, Mendiratta RG, Alam MI. Dielectric properties of NiZn ferrites prepared by the citrate precursor method. Materials Science and Engineering B. 1999;**60**:156

[12] Rao SK, Abhinav EM, Jaison D, Sundararaj A, Santhiya M, Althaf R, et al. Investigation of room temperature multi-functional properties of Nd doped mullite $Bi_2Fe_4O_9$. Vacuum. 2020;**172**:109109

[13] Ameer S, Jindal K, Tomar M, Jha PK, Gupta V. Insight into electronic, magnetic, and optical properties of magnetically ordered $Bi_2Fe_4O_9$. Journal of Magnetism and Magnetic Materials. 2019;**475**:695

[14] Ameer S, Jindal K, Tomar M, Jha PK, Gupta V. Tunable electronic and magnetic properties of 3d transition metal doped $Bi_2Fe_4O_9$. Journal of Magnetism and Magnetic Materials. 2020;**509**:166893

[15] Pooladi M, Shokrollahi H, Lavasani SANH, Yang H. Investigation of the structural, magnetic, and dielectric properties of Mn-doped $Bi_2Fe_4O_9$ produced by reverse chemical co-precipitation. Materials Chemistry and Physics. 2019;**229**:39-48

[16] Moghadam AK, Mirzaee O, Shokrollahi H, Lavasani SANH. Magnetic and morphological characterization of bulk $Bi_2Fe_4O_9$ derived by reverse chemical co-precipitation: A comparative study of different sintering

methods. Ceramics International. 2019;**45**:8087-8094

[17] Zdorovets MV, Kozlovskiy AL. Study of phase transformations in Co/ $CoCo_2O_4$ nanowires. Journal of Alloys and Compounds. 2020;**815**:152450

[18] Kozlovskiy AL, Kenzhina IE, Zdorovets MV. FeCo– Fe_2CoO_4/Co_3O_4 nanocomposites: Phase transformations as a result of thermal annealing and practical application in catalysis. Ceramics International. 2020;**46**:10262-10269

[19] Rusakov VS, Kadyzhanov KK, Kozlovskiy AL, Fadeev MS, Zdorovets MV. Phase transformations as a result of thermal annealing of nanocomposite Fe–Ni/Fe–Ni–O particles. Ceramics International. 2020;**46**:1586-1595

[20] Kozlovskiy A, Kenzhina I, Zdorovets M. Synthesis, phase composition and magnetic properties of double perovskites of A(FeM)O4-x type (A=Ce; M=Ti). Ceramics International. 2019;**45**:8669-8676

[21] Kozlovskiy A, Dukenbayevd K, Kenzhinaa I, Tosid D, Zdorovetsa M. Dynamics of changes in structural properties of AlN ceramics after Xe+22 ion irradiation. Vacuum. 2018;**155**:412

[22] Rodriguez-Carvajal J. Recent advances in magnetic structure determination by neutron powder diffraction. Physica B. 1993;**192**:55

[23] Zdorovets MV, Kozlovskiy AL. The effect of lithium doping on the ferroelectric properties of LST ceramics. Ceramics International. 2020;**45**:14548-14557

[24] Cai D, Du D, Yu S, Cheng J. Oriented growth of $Bi_2Fe_4O_9$ crystal and its photocatalytic activity. Procedia Engineering. 2012;**27**:577-582

[25] Tian ZM, Qiu Y, Yuan SL, Wu MS, Huo SX, Duan HN. Enhanced multiferroic properties in Ti-doped $Bi_2Fe_4O_9$ ceramics. Journal of Applied Physics. 2010;**108**:064110

[26] Bell AT. The impact of nanoscience on heterogeneous catalysis. Science. 2003;**299**:1688

[27] Liu ZS, Wu BT, Yin DG, Zhu YB, Wang LG. Enhanced photocatalytic activity in Al-substituted $Bi_2Fe_4O_9$ submicrocrystals. Journal of Materials Science. 2012;**47**:6777-6783

[28] Friedrich A, Biehler J, Morgenroth W, Wiehl L, Winkler B, Hanfland M, et al. High-pressure phase transition of $Bi_2Fe_4O_9$. Journal of Physics: Condensed Matter. 2012;**24**:145401

[29] Massey MJ, Baier U, Merlin R, Weber WH. Effects of pressure and isotopic substitution on the Raman spectrum of α-Fe_2O_3: Identification of two-magnon scattering. Physical Review B. 1990;**41**:7822

[30] Litvinchuk AP, Börjesson L, Thomsen C, Chu CW. Magnetic excitations in $PrBa_2Cu_4O_8$ explored by raman scattering. Physica Status Solidi B: Basic Solid State Physics. 1999;**215**:507

[31] Holmlund J, Knee CS, Andreasson J, Granath M, Litvinchuk AP, Börjesson L. Two-magnon Raman scattering from the Cu_3O_4 layers in (Sr_2, Ba_2) $Cu_3O_4Cl_2$. Physical Review B. 2009;**79**:085109

[32] White RM, Nemanich RJ, Herring C. Light scattering from magnetic excitations in orthoferrites. Physical Review B. 1982;**25**:1822

[33] Park YA, Song KM, Hur N. Frequency-Dependent dielectric anomalies in magnetic oxides. Journal of the Korean Physical Society. 2008;**53**:3356

[34] Shukla A, Choudhary RNP. Study of electrical properties of La^{3+}/Mn^{4+}-modified $PbTiO_3$ nanoceramics. Journal of Materials Science. 2012;**47**:5074

[35] Yamazaki Y, Satou M. High frequency conductivity in Cobalt-Iron ferrite. Japanese Journal of Applied Physics. 1973;**12**:998-1000

[36] Jonscher AK. Dielectric relaxation in solids. Journal of Physics D: Applied Physics. 1996;**32**:R57

[37] El Hiti MA. AC electrical conductivity of Ni-Mg ferrites. Journal of Physics D: Applied Physics. 1996;**29**:501-505

[38] Panda A, Govindaraj R, Amarendra G. Magneto dielectric coupling in $Bi_2Fe_4O_9$. Physica B: Condensed Matter. 2019;**570**:206-208

[39] Yuan X, Shi L, Zhao J, Zhou S, Guo J. Tunability of magnetization and bandgap in mullite-type $Bi_2Fe_4O_9$ ceramics through non-magnetic ions. Scripta Materialia. 2018;**146**:55-59

[40] Liu T, Xu Y, Zeng C. Synthesis of $Bi_2Fe_4O_9$ via PVA sol–gel route. Materials Science and Engineering B. 2011;**176**:535-539

[41] Dormann JL, Nogues M. Magnetic structures in substituted ferrites. Journal of Physics: Condensed Matter. 1990;**2**:1223

Chapter 2

Electric Field-Induced Magnetization Reversal of Multiferroic Nanomagnet

Jiahao Liu and Liang Fang

Abstract

Using the inverse piezoelectric effect and inverse magnetostrictive effect in a multiferroic heterojunction, an electric field is able to control the magnetization switching of a uniaxial nanomagnet. Compared with traditional spintronic devices based on magnetic field, multiferroic nanomagnet devices have the advantages of ultra-low consumption and high radiation resistance, showing great application potential in modern high-integrated circuits and military electronic systems. However, the difficulties of electric field control of complete magnetization reversal of the nanomagnet and nanomagnet arrays in a nanomagnetic logic gate still restrict the developments of multiferroic nanomagnet device. In this chapter, the uniaxial nanomagnets in multiferroic heterojunctions are mainly discussed. The two core problems of the electric field control of nanomagnets and nanomagnetic logic gate are well solved.

Keywords: multiferroics, magnetization switching, spintronics, nanomagnet, straintronics

1. Introduction

Using the inverse piezoelectric effect and inverse magnetostrictive effect in a multiferroic heterojunction, an electric field is able to control the magnetization switching of a uniaxial nanomagnet. **Figure 1** is a multiferroic heterojunction structure, that is, a two-layer magnetoelectric composite structural system, which is formed by magnetoelastic coupling of a magnetostrictive layer and a piezoelectric layer. The electric field-controlled nanomagnet in a multiferroic heterojunction essentially uses multi-field coupling of "electric-stress-magnetic." Applying a small voltage to the piezoelectric layer, the piezoelectric layer will produce uniaxial strain, which is transformed into a stress applying on the magnetostrictive layer by magnetoelastic coupling, causing the magnetization direction of the magnetostrictive layer to rotate perpendicular to the stress. If the magnetostrictive layer is constructed as a uniaxially shaped nanomagnet, the strain will reverse the magnetization direction of the nanomagnet to a logically "NULL" state, pointing to the hard magnetization axis, which is, the short axis direction. At this time, if the voltage is released (stress revocation), the nanomagnet will flip to a certain stable logic state (original logic state or opposite logic state). This magnetic logic device based on the regulation mode of the multiferroic heterojunction magnetoelectric effect is called

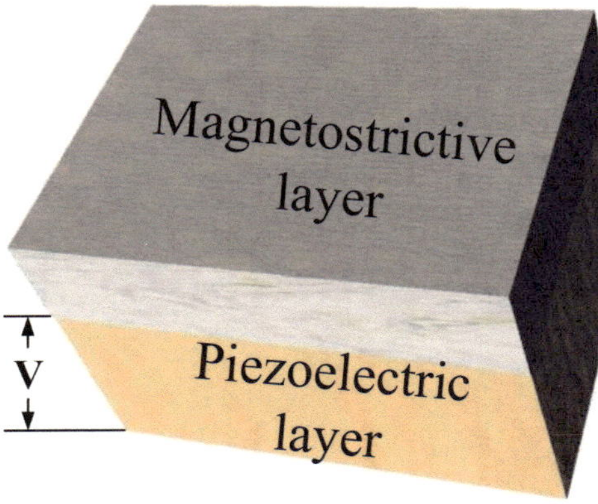

Figure 1.
Multiferroic heterojunction structure.

a multiferroic nanomagnet device. Multiferroic nanomagnet device is one of the most competitive spintronic devices due to its low energy consumption and high thermal stability [1]. It represents Boolean logic "0" and "1" in the magnetization directions along the long axis of a uniaxial nanomagnet [2] and can be corresponding to different states in the magnetic tunnel junction [3].

2. Voltage pulse-induced magnetization switching

The key to using the stress generated by the electric field to control the magnetization switching of the multiferroic nanomagnet is that the stress anisotropy must be larger than the shape anisotropy of the nanomagnet. An effective method to reduce the required stress anisotropy is to break the symmetry of the shape of the nanomagnet by slightly tilting the long axis of the nanomagnet to the direction of stress application. However, the effect of the tilt angle on the magnetization reversal of the nanomagnet is still inconclusive.

2.1 Model

Figure 2 presents the voltage-controlled multiferroic heterostructure. The red arrow indicates the direction of magnetization. The polar angle (out-of-plane) and the azimuth angle (in-plane) of the magnetization are θ and φ, respectively. Note that the magnet is at an angle to the direction of the electrodes.

The magnetization dynamic of a single elliptical nanomagnet meets the Landau-Lifshitz-Gilbert Eq. (5):

$$\frac{\mathrm{d}\vec{M}}{\mathrm{d}t} = -\gamma\vec{M} \times \overrightarrow{H_{eff}} - \frac{\alpha\gamma}{M_s}\left[\vec{M} \times \left(\vec{M} \times \overrightarrow{H_{eff}}\right)\right] \tag{1}$$

where α is the damping coefficient, $\vec{M}$ is the magnetic moment vector of the nanomagnet, M_s is the saturation magnetization, γ is the return ratio, and [5]:

$$\overrightarrow{H_{eff}} = -\frac{1}{\mu_0 V}\frac{\mathrm{d}E}{\mathrm{d}\vec{M}} \tag{2}$$

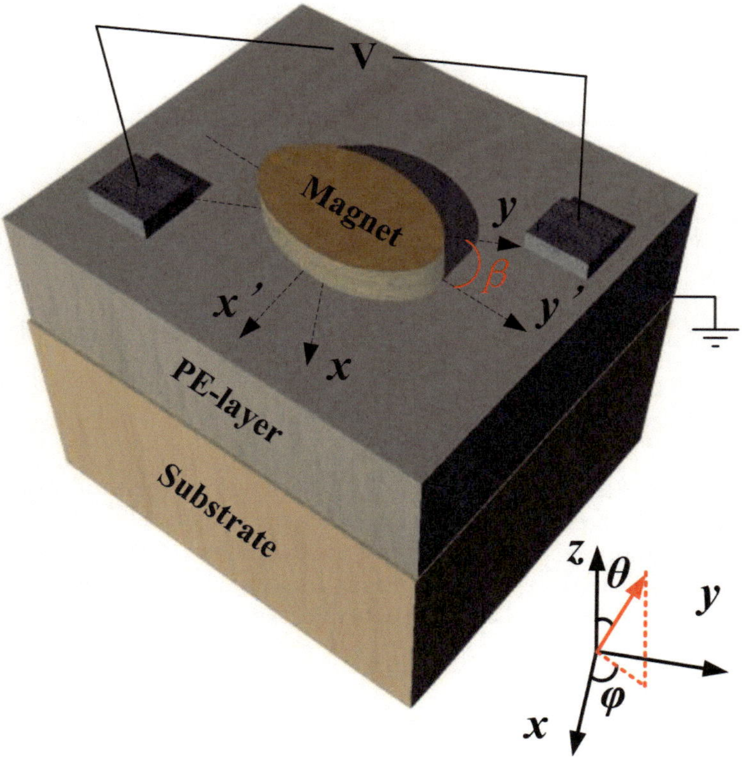

Figure 2.
Stress-regulated multiferroic tilted nanomagnet device [4].

is the effective field generated by a variety of energies (shape anisotropy energy, stress anisotropy energy, Zeeman energy, and thermal fluctuations), where $\mu_0 = 4\pi \times 10^{-7}$ is the vacuum permeability and V is the volume of each element. The stress is applied at the y direction, and the total energy E_{total} is the sum of demagnetization energy, exchange energy, shape anisotropy energy, stress anisotropy energy, and energy dissipation:

$$E_{\text{total}} = E_{\text{demagnetization}} + E_{\text{exchange}} + E_{\text{shape-anisotropy}} + E_{\text{stress-anisotropy}} + E_{\text{dissipation}} \tag{3}$$

For Terfenol-D as the magnetic material, the crystal anisotropy energy of is small, and thus is ignored in the calculation of the total energy. The exchange energy can also be neglected in the single domain particles of 100 nm × 50 nm × 20 nm [6]. The shape anisotropy energy of the nanomagnet can be written as [7]:

$$E_{\text{shape-anisotropy}} = \int -(\mu_0/2)\, \vec{\boldsymbol{M}} \cdot \vec{\boldsymbol{H}}_M \, \mathrm{d}V \tag{4}$$

where $\vec{\boldsymbol{M}}$ is the magnetic moment vector of the nanomagnet and $\vec{\boldsymbol{H}}_M$ is the shape anisotropy energy field, which can be expressed as [7]:

$$\vec{\boldsymbol{H}}_M = -N_{dx}M_x\hat{i} - N_{dy}M_y\hat{j} - N_{dz}M_z\hat{k} \tag{5}$$

$$E_{\text{shape-anisotropy}} = -(\mu_0/2)\left(M_s^2 V\right)\left[N_{dx}\cos^2\theta\sin^2\varphi + N_{dy}\sin^2\theta\sin^2\varphi + N_{dz}\cos^2\varphi\right] \tag{6}$$

where N_d is the demagnetization factor. For elliptical shaped nanomagnets, the demagnetization factors N_{dx}, N_{dy}, and N_{dz} can be calculated through [7]:

$$N_{dx} = \frac{\pi}{4}\left(\frac{th}{a}\right)\left[1 + \frac{5}{4}\left(\frac{a-b}{a}\right) + \frac{21}{16}\left(\frac{a-b}{a}\right)^2\right] \tag{7}$$

$$N_{dy} = \frac{\pi}{4}\left(\frac{th}{a}\right)\left[1 - \frac{1}{4}\left(\frac{a-b}{a}\right) - \frac{3}{16}\left(\frac{a-b}{a}\right)^2\right] \tag{8}$$

$$\mathrm{N_{dz}} = 1 - \mathrm{N_{dx}} - \mathrm{N_{dy}} \tag{9}$$

where a is the length of the long axis, b is the length of the short axis, and th is the thickness of the nanomagnet. For a nanomagnet whose tilt angle is β, as shown in **Figure 2**, the short axis and long axis of the nanomagnet rotate clockwise from the x axis and y axis to the x' axis and y' axis, respectively, and z' axis (not shown) is still at vertical direction. The shape anisotropy field components in the new coordinate system are:

$$h'_{\text{shape-anisotropy\_xx}} = -M_s N_{dx}\cos(\varphi+\beta)\sin\theta \tag{10}$$

$$h'_{\text{shape-anisotropy\_yy}} = -M_s N_{dy}\sin(\varphi+\beta)\sin\theta \tag{11}$$

$$h'_{\text{shape-anisotropy\_zz}} = -M_s N_{dz}\cos\theta \tag{12}$$

By the coordinate rotation conversion of (10)–(12), the field components of shape anisotropy of the tilted nanomagnet on the original coordinate axes become:

$$h_{\text{shape-anisotropy\_xx}} = -M_s\left(N_{dx}\cos(\varphi+\beta)\cos\beta + N_{dy}\sin(\varphi+\beta)\sin\beta\right)\sin\theta \tag{13}$$

$$h_{\text{shape-anisotropy\_yy}} = -M_s\left(-N_{dx}\cos(\varphi+\beta)\sin\beta + N_{dy}\sin(\varphi+\beta)\cos\beta\right)\sin\theta \tag{14}$$

$$h_{\text{shape-anisotropy\_zz}} = -M_s N_{dz}\cos\theta \tag{15}$$

The stress anisotropy energy of the nanomagnet is given by [7]:

$$E_{\text{stress-anisotropy}} = -\frac{3}{2}\lambda_s\sigma V\sin^2\theta\sin^2\varphi \tag{16}$$

where $3\lambda_s/2$ is the saturation magnetostriction and the stress σ is considered negative for compression and positive for tension. The stress is applied in the y direction, so there is only a field component in the y axis direction [8]:

$$h_{\text{stress-anisotropy\_yy}} = (3\lambda_s/M_s\mu_0)\sigma\sin\theta\sin\varphi \tag{17}$$

Considering the thermal fluctuations, the effect of random thermal fluctuations can be described by a random thermal field [9]:

$$h(t) = \sqrt{\frac{2\alpha kTf}{\gamma\mu_0 M_s V}}G_{(0,1)}(t) \tag{18}$$

where k = 1.38 × 10^{-23} J/K is the Boltzmann constant, T = 300 K is the room temperature, f = 1 GHz is the frequency of thermal noise oscillations, and $G_{(0,1)}$ represents a Gaussian function with a mean of 0 and a variance of 1. By combining the above functions, the effective field $\overrightarrow{H_{eff}}$ can be obtained. A bias field can also be involved in the x direction if it is required. The components of each coordinate axis are:

$$h_x = -M_s\left(N_{dx}\cos(\varphi+\beta)\cos\beta + N_{dy}\cos(\varphi+\beta)\sin\beta\right)\sin\theta + \sqrt{\frac{2\alpha kTf}{\gamma\mu_0 M_s V}}G_{(0,1)}(t) + h_{bias} \quad (19)$$

$$h_y = -M_s\left(N_{dx}\cos(\varphi+\beta)\sin\beta + N_{dy}\sin(\varphi+\beta)\cos\beta\right)\sin\theta + (3\lambda_s/M_s\mu_0)\sigma\sin\theta\sin\varphi + \sqrt{\frac{2\alpha kTf}{\gamma\mu_0 M_s V}}G_{(0,1)}(t) \quad (20)$$

$$h_z = -M_s N_{dz}\cos\theta + \sqrt{\frac{2\alpha kTf}{\gamma\mu_0 M_s V}}G_{(0,1)}(t) \quad (21)$$

2.2 Results and discussions

Biswas et al. used two pairs of electrodes to control the nanomagnet in the experiment to achieve a reliable 180° switching [10, 11]. However, since the two pairs of electrodes have to be operated in sequence, the nanomagnet needs a longer switching time. Fashami used a timed pulse to switch the nanomagnet by 180°, which is error-free and dissipates arbitrarily small energy [12]. However, in this scheme, a hard magnet is essential to break the energy symmetry, and a long switching time is required. Recently, a method of 180° switching has been proposed, in which a repeatable 180° nanomagnet switching was induced by voltage pulses. By setting suitable amplitude, width, and period of the voltage pulse, it is possible to achieve repeatable 180° switchings without a magnetic field [13, 14]. However, although this solution can achieve repeatable magnetic switching, the first switching requires a large start-up time, making the first switching time much longer [15, 16]. In magnetic storage and logic application, the first switching is most often needed. More importantly, these studies did not consider the thermal fluctuations, which play an important role in the switching of the nanomagnet. In conclusion, fast switching of nanomagnets at room temperature is still a challenge for straintronics in the application of logic storage and computing. This section introduces a fast switching method of nanomagnets at room temperature. The structure is shown in **Figure 1** of the previous section. The authors use OOMMF software to simulate and study the switching of nanomagnets.

The authors chose PMN-PT ($Pb(Mg_{1/3}Nb_{2/3})O_3$-$PbTiO_3$) as the piezoelectric layer material to use its higher piezoelectric coefficient [17, 18]. And for the magnetic material, the authors chose Terfenol-D ($Tb_{0.7}Dy_{0.3}Fe_2$), because the magnetocrystalline anisotropy can be smaller [19]. The parameters are shown in **Table 1**.

Since (Object Oriented Micromagnetic Framework) software [20] cannot directly set the stress anisotropy energy, the authors use the uniaxial anisotropy energy acting in the direction of (−cosβ sinβ 0) for replace. Accordingly [6]:

$$K = -\frac{3}{2}\lambda_s\sigma V \quad (22)$$

Symbol	Quantity	Value
Y	young's modulus	8×10^{10} Pa
$3\lambda_s$ 2	saturated magnetostriction	6×10^{-4}
α	damping coefficient	0.1
K	magnetocrystalline anisotropy constant	0
M_s	saturated magnetization	8×10^{5} A m
a	length of long axis	100 nm
b	length of short axis	50 nm
th	thickness of magnetic layer	20 nm
t_p	thickness of piezoelectric layer	400 nm
υ	Poisson's ratio	0.3

Table 1.
Parameters of multiferroic heterostructure.

The size of nanomagnet is 51 nm × 102 nm × 21 nm. The selection of large aspect ratio and thickness can reduce C-shaped and eddy vortex errors [21]. The mesh size of OOMMF is 3 nm × 3 nm × 3 nm. Magnetization toward up and down is defined as logic "1" and "0," respectively. The initial state of the nanomagnet is assumed as logic "1."

Before studying voltage pulse-induced 180° switching, the first step is to ensure that the magnetization direction of the nanomagnet is able to rotate by more than 90° (below x axis). **Figure 3** shows that minimum stress is required for the nanomagnet to rotate by more than 90° when the stress is applied in different directions ($0 < \beta < 10°$) [22]. A small β can reduce the required stress, which makes it easier for the nanomagnet to rotate by more than 90°. However, as β increases, the required stress also increases. This is because the stress tends to make the magnetization direction perpendicular to the axis of the electrodes pair, i.e., to flip to the x' axis. When β is larger, the x' axis will also make a larger deflection angle with the x axis. This makes nanomagnet difficult to rotate by more than 90°. Even so, when $0 < \beta < 7°$, the required stress is less than the scheme with the electrodes' pair axis along the long axis of the nanomagnet ($\beta = 0$).

In the second step, optimal voltage pulse should be set to make the switching time as short as possible. The authors apply a stress of 100 MPa to the nanomagnet (voltage pulse peak of 225 mVs), which is sufficient for the nanomagnet to rotate by more than 90°. **Figure 4** shows the optimal waveform setting and dynamic

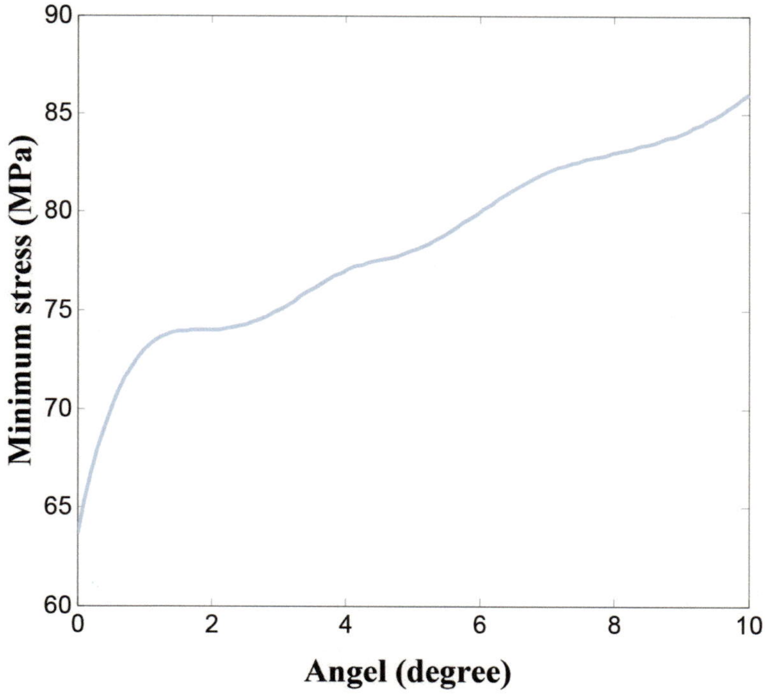

Figure 3.
Minimum stress required for the nanomagnet to rotate by more than 90° with different tilt angles β of electrodes' pair axis [22].

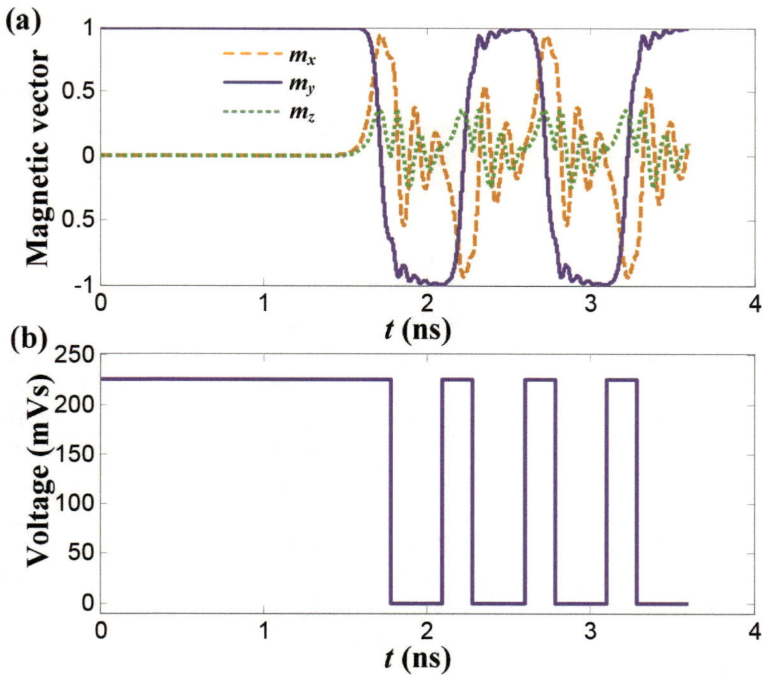

Figure 4.
180° switching with the stress electrodes' pair axis along the long axis of the nanomagnet. (a) Dynamic magnetization. (b) Voltage pulse waveform.

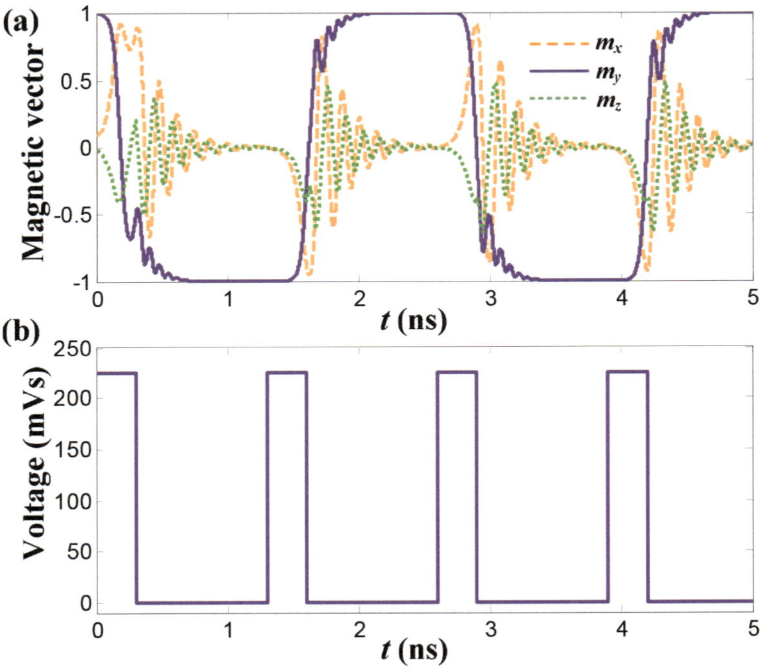

Figure 5.
Repeated 180° switching with β = 5° under room temperature. (a) Dynamic magnetization. (b) Voltage pulse waveform.

magnetization of the repeatable 180° switchings in the nanomagnet, when the electrodes' pair axis is aligned with the long axis of the nanomagnet ($\beta = 0$). **Figure 5(a)** is the dynamic magnetization of repeatable 180° switchings in the nanomagnet. It can be seen from the inset that the nanomagnet requires the voltage to be applied for a long period of time (1.6 ns) before it can enter the switching cycle. This is because the nanomagnet has equal probability of reaching either orientation when the stress is applying along the long axis. Therefore, the nanomagnet will enter a magnetization direction selection period before it can be flipped. This start-up time greatly increases the first switching time of the nanomagnet. **Figure 5(b)** shows the optimal voltage pulse waveform for the nanomagnet switching. The minimum start-up time of the nanomagnet $t_{\text{start-up}} = 1.600$ ns, the minimum voltage pulse width $t_{\text{width}} = 0.180$ ns, and the minimum pulse interval time $t_{\text{interval}} = 0.320$ ns. The minimum switching period of the nanomagnet is the sum of the minimum voltage pulse width and the minimum interval time, $T = t_{\text{width}} + t_{\text{interval}} = 0.500$ ns, and the maximum switching frequency $f = 1/T = 2.000$ GHz. The time that the nanomagnet completes the initial switching is the sum of the minimum start-up time and the minimum switching period: $t_{\text{initial}} = t_{\text{starting}} + T = 2.070$ ns.

If the electrodes' pair axis is not aligned with the long axis of the nanomagnet, but is tilted by a small angle β, the nanomagnet will have a tendency to select where to flip. For $\beta > 0$, nanomagnets tend to flip clockwise. This allows the nanomagnet to require no start-up time during the first switching, greatly increasing the efficiency of the initial switching.

Figure 6 shows the dynamic magnetization of the switchings and optimal voltage pulse waveform when $\beta = 5°$. The nanomagnet has no start-up time and directly enters the switching cycle. The minimum voltage pulse width $t_{\text{width}} = 0.162$ ns, and the minimum pulse interval time $t_{\text{interval}} = 0.312$ ns. Therefore, the minimum switching period of the nanomagnet is the sum of the minimum voltage pulse width

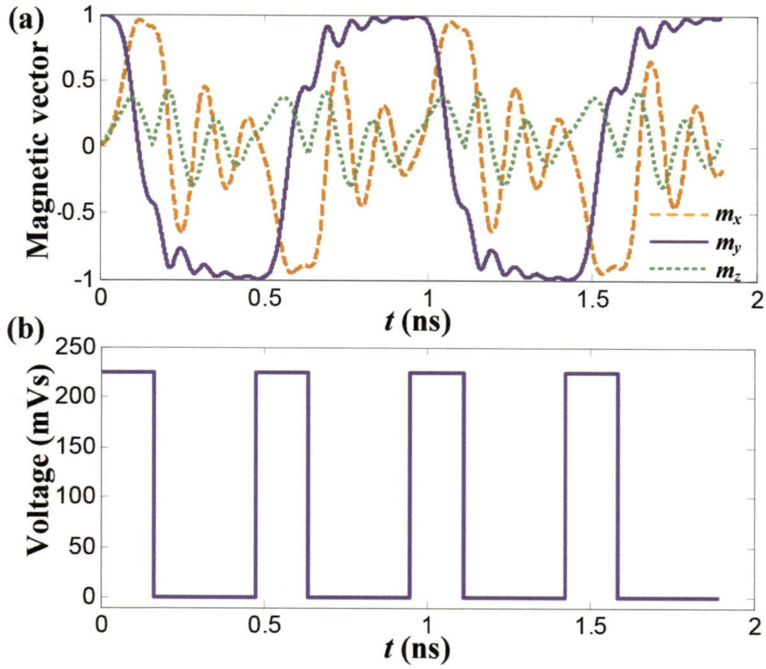

Figure 6.
Fast 180° switching with β = 5°. (a) Dynamic magnetization. (b) Voltage pulse waveform.

and the minimum interval time, $T = t_{\text{width}} + t_{\text{interval}} = 0.447$ ns, and the maximum switching frequency $f = 1/T = 2.110$ GHz. The time that the nanomagnet completes the initial switching is the same as the minimum switching cycle: $t_{\text{initial}} = T = 0.474$ ns. This is only about 1/5 of **Figure 4**.

The authors continue to calculate the minimum time and maximum switching frequency for the nanomagnet to complete the 180° switching. The voltage pulse peak is controlled to be a constant 225 mVs.

As shown in **Figure 7**, since there is no start-up time, the minimum initial switching time of the nanomagnet with $\beta > 0$ is significantly smaller than that of the nanomagnet with $\beta = 0$. The minimum pulse width decreases as β increases. For $\beta > 6°$, although the minimum pulse width continues to decrease as β increases, the minimum interval increases in the meanwhile. When $4° < \beta < 9°$, the minimum total initial switching time is small and the maximum switching frequency is also larger than that of **Figure 5** ($\beta = 0$). Based on the above factors, β should be chosen to be around 5°. So nanomagnets will have less required stress, larger switching frequency, and shorter initial switching time.

Although voltage pulse-induced magnetization switching is very energy efficient, the possibility of operating at room temperature remains to be discussed, which plays an important role in the switching. In this section, the switching of the nanomagnet at room temperature (300 K) is calculated. Since OOMMF software could be computationally expensive and time-consuming to simulate the switching at room temperature, the authors use the mathematical stress model to calculate the switching of the nanomagnet at room temperature.

Firstly, the authors apply a stress of 100 MPa to the electrodes and observe the dynamic magnetization of the nanomagnet. The magnetization rotates by more than 90° at 0.1844–0.3470 ns and is most close to logic "0" at 0.2574 ns, meaning that $t_{\text{width}} = 0.2574$ ns. If the stress is removed at 0.2574 ns, the nanomagnet will flip to logic "0" at 0.7427 ns, meaning that $t_{\text{interval}} = 0.8856$ ns. Obviously, at room temperature, both t_{width} and t_{interval} are much larger than that at 0 K.

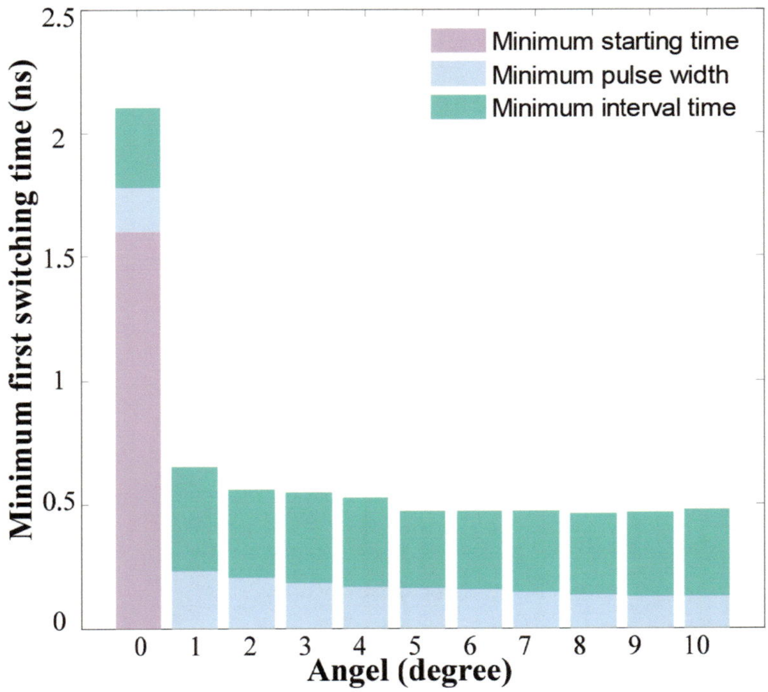

Figure 7.
Minimum switching times of the nanomagnet changes with the tilt angles of the electrode pair axis.

Secondly, the authors try to control the switching by voltage pulse of t_{width} = 0.2574 ns and $t_{interval}$ = 0.8856 ns at room temperature. Unfortunately, the nanomagnet succeeds to be switched to logic "0," but never return back to logic "1" again. Under the influence of thermal fluctuations, the nanomagnet needs to remain in a stable logic state for a longer period of time before it can be switched again.

Thirdly, t_{width} = 0.3 ns (it can be chosen from 0.1844 to 0.3470 ns) and $t_{interval}$ = 1 ns are given to gain repeated switchings. As shown in **Figure 5**, the nanomagnet converts back and forth between two logic states at room temperature successfully.

The switching cycle is 1.3 ns. One thing that must be pointed out is that the simulations assume the ideal voltage pulse waveform. The effects of the rising and falling edges of the actual voltage pulse are not considered. Besides, through the calculation of the model, $t_{interval}$ can be long enough but t_{width} can be only chosen from 0.1844 to 0.3470 ns. Actually, pulse width needed for 180° switching will not be a constant in the presence of thermal noise. The pulse width error of less than 0.2 ns will be a challenge under room temperature, which is the disadvantage of this scheme. These will need to be further studied in the subsequent experimental work.

Due to the symmetry, setting the initial logic as "0" or setting the electrodes' pair axis, a clockwise deflection will get the same result, which is not described in this paper for clarity.

2.3 Conclusion

The efficient 180° switching of the magnetization direction of the nanomagnet is the key to straintronic devices in the application of magnetic storage and logic. The voltage pulse-induced repeatable 180° switching is a fast and low energy consumption scheme, but the initial switching requires a large start-up time and thermal fluctuation is a great challenge. This method overcomes the start-up time of the initial switching by rotating the stress electrodes' pair axis by a small angle from the long axis of the nanomagnet. Using OOMMF software for simulation, the optimal voltage pulse waveform to control the 180° switchings of the nanomagnet is calculated, and the influence of electrodes' pair axis tilt angle β is studied. The results show that when the tilt angle β is about 5°, the nanomagnet has lower switching frequency, shorter initial switching time, and less required stress. Repeated switching at room temperature is calculated by mathematical model. The switching time is longer under the influence of thermal fluctuations. These findings will provide possible guidance for straintronic devices in the application of magnetic storage and logic.

3. Electric control of nanomagnetic logic gate

The previous section introduced the electric field regulation of a single nanomagnet, and this section will continue to discuss the electric field control method for nanomagnet arrays. Information transmission and calculations in nanomagnetic logic rely on the control of nanomagnet array. The problem of efficient information transmission is well solved [6]. However, electric-controlled magnetic logic gate is still a major challenge. Imre et al. used five single-axis nanomagnets to build a majority logic gate [2], which made nanomagnetic logic possible. However, this logic gate requires multiple clock controls to ensure correct logic calculations. Gypens et al. used 19 dipole-coupled uniaxial nanomagnets to form a stable system and built a NAND (NOR) logic gate that can be accurately calculated [23]. However, this solution requires more nanomagnets, which increases the NML area. Roy uses a multiiron material to propose an ultra-low-energy NAND (OR) logic gate based on a magnetic tunnel junction [24]. However, this logic gate design requires casting multiple layers of materials, which increases the difficulty of manufacturing. Niemier et al. put forward a long axis tilted nanomagnet structure by using an edge-slanted nanomagnet and designed dual-input AND/OR logic gates based on it. Most studies now use this type of edge-slanted nanomagnet to achieve long axis tilted nanomagnet structures. However, there are three defects in edge-slanted nanomagnets: (1) This type of nanomagnet requires a larger size, thus increasing the NML space and introducing clock errors of the C-shape and eddy current that easily occur in large-sized nanomagnets. (2) Complex calculations caused by the irregular shape are inevitable. (3) More importantly, the irregular shape of nanomagnet increases the requirements of fabrication process.

From the above perspective, a more effective and more reliable design of basic magnetic logic gates is required to be proposed. The design should address two key issues: (1) how to eliminate C-shaped and eddy current clock errors and (2) how to reduce the complexities of calculations and fabrication process.

3.1 Design and analysis

In the previous section, the long axis tilted nanomagnet is introduced. As shown in **Figure 8(a)**, the long axis and short axis of the nanomagnet rotate from the x axis and y axis to the x' axis and y' axis, respectively. If the tilt angle that long axis makes with the direction of the electrodes is β, the included angles between long axis and the clock will be a larger one ($90° + \beta$) and a smaller one ($90° - \beta$). When driven by no other energy, the nanomagnet will flip toward the smaller angle after the stress is released. This is because the nanomagnet has higher anisotropy along the clock than that along the long axis and will spontaneously flip to the shape anisotropy potential

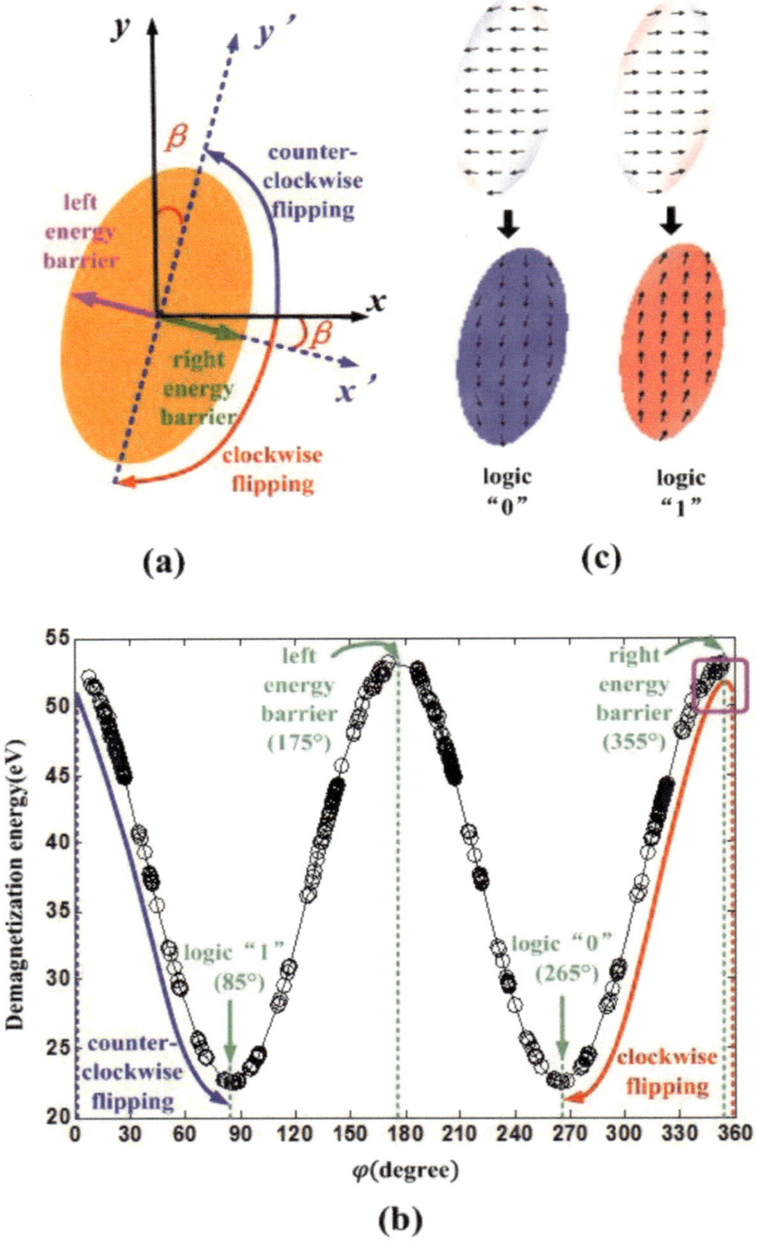

Figure 8.
(a) The nanomagnet is rotated clockwise by a small angle β. (b) the demagnetization energy is calculated as a function of φ. (c) the preferred magnetization is simulated by OOMMF.

well of the long axis. However, in the course of flipping toward the larger angle (90° + β), it is necessary to cross the shape anisotropy barrier of the hard axis. As a consequence, the nanomagnet tends to flip toward the smaller angle (90° − β) without the need of crossing the shape anisotropy barrier of the hard axis.

As shown in **Figure 8(b)**, for a nanomagnet with a tilt angle β = 5°, the demagnetization energy is calculated by OOMMF, as a function of φ. For the parameters, the authors have assumed a space size of 80 nm × 100 nm × 20 nm, mesh size of 2 nm × 2 nm × 2 nm, magnet dimensions of 50 nm × 100 nm × 20 nm, saturation magnetization of 800 kA/m, Gilbert damping constant of 0.5, and zero magnetocrystalline anisotropy. The high aspect ratio (2:1) and the small tilt angle

(β = 5°) of the nanomagnet are set to eliminate the C-shaped and eddy current clock errors. As shown in the inset, the demagnetization energy curve of the tilted nanomagnet is shifted 5° to the left, where logic "1" and "0" correspond to 85° and 265°, respectively, while "NULL" (high energy state) states correspond to 175° and 355°. If the initial clock of the nanomagnet is pointing right (φ = 0 or 360°), after the stress is released, the tilted nanomagnet will flip counterclockwise to the side that is at a smaller angle to the long axis, which is the *+y'* direction (85°). This is because the nanomagnet needs to cross the right shape anisotropy barrier of the hard axis (see the purple box shown in **Figure 8(b)**) when turning clockwise to the *−y'* direction (265°), whereas when turning counterclockwise, it is not necessary to cross the barrier. Thus the nanomagnet will flip counterclockwise to the *+y'* direction, yielding logic "1." **Figure 8(c)** gives the OOMMF simulations of the preferred magnetization of the nanomagnet with initial clock pointing left or right. As shown in the inset, if the initial state is pointing left, the tilted nanomagnet will rotate counterclockwise to logic "0," whereas if the initial state is pointing right, it will rotate counterclockwise to logic "1."

Based on the preferred magnetization of tilted nanomagnet, a design of dual-input AND/OR magnetic logic gates is proposed, as shown in **Figure 9**. This design is composed of two input nanomagnets A and B, as well as one output tilted nanomagnet Out (clinched 5° clockwise), interacting via ferromagnetic coupling. The magnetization direction of the magnet Out is influenced by the ferromagnetic coupling of the input magnets A and B as well as its own preferred magnetization. As shown in **Figure 9(a)**, if the initial state is pointing left, the nanomagnet Out tends to flip to logic "0." As a consequence, when the inputs A and B are "01," "00," or "10," the output magnet rotates counterclockwise to logic "0," whereas when the inputs A and B are both "1," the output magnet rotates clockwise to logic "1," thereby yielding AND logic. If the initial state is pointing right, as shown in **Figure 9(b)**, the nanomagnet Out tends to flip to logic "1," so when the input magnets A and B are "01," "11," or "10," the output magnet rotates

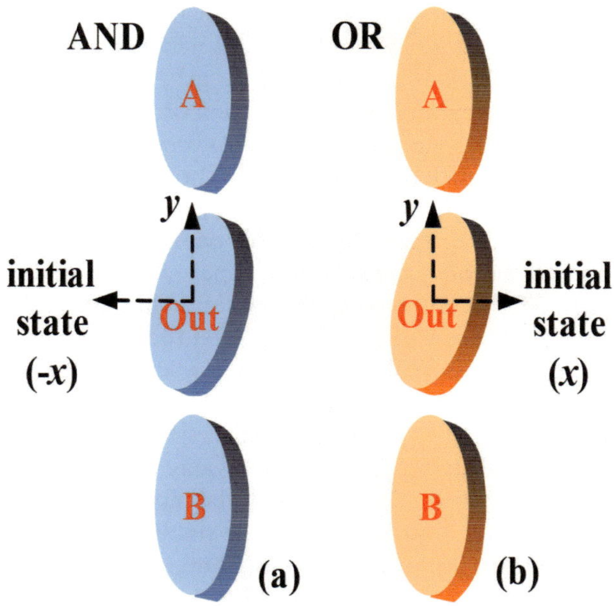

Figure 9.
(a) and (b) show the design of (A) AND logic gate and (B) OR logic gate based on tilted nanomagnet. The initial magnetization of magnet Out is pointing left in (a) and right in (b).

counterclockwise to logic "1," whereas when inputs A and B are both "0," the output magnet rotates clockwise to logic "0," yielding OR logic.

For magnet Out, whose magnetization is interacted by inputs A and B, the dipole–dipole interaction energy writes [7]:

$$E_{\text{dipole}} = \frac{\mu_0 M_s^2 V^2}{4\pi R^3}\left[\begin{pmatrix} -2\cos\varphi_A \sin\theta_A \cos\varphi \sin\theta \\ +\sin\varphi_A \sin\theta_A \sin\varphi \sin\theta + \cos\theta_A \cos\theta \end{pmatrix} + \begin{pmatrix} -2\cos\varphi_B \sin\theta_B \cos\varphi \sin\theta \\ +\sin\varphi_B \sin\theta_B \sin\varphi \sin\theta + \cos\theta_B \cos\theta \end{pmatrix}\right] \tag{23}$$

where R is the separation between the centers of neighbor nanomagnets and the magnetization angles of the input magnets are labeled with subscripts A and B.

3.2 Results and discussions

Only OR logic gate is discussed in this section. For shape symmetry, the results will be same for AND logic gate; on account of which, it is not discussed here for clarity. In order to obtain OR logic gate, an initial clock pointing right is necessary. However, whether the clock direction is pointing left or right cannot be controlled simply by the stress. The magnetization vector only tends to be perpendicular to where the stress is applied. Fortunately, for the nanomagnet tilted clockwise by 5°, the direction of initial clock will be determined by the initial magnetization direction of the nanomagnet. As mentioned in Section II, there is no need of crossing the hard axis barrier for the magnet when flipping clockwise. As a consequence, a nanomagnet whose initial state is logic "1" (φ = 90°) tends to flip clockwise under the stress applied in the y direction. It is worth mentioning that if it is not possible to know the initial state of the tilted nanomagnet, a clock pointing right can be obtained by adding a biasing magnetic field pointing right (a stress of 45 MPa and a bias magnetic field of 500 Oe).

The authors assume that the initial state of the nanomagnet Out is logic "1" (φ = 90°, θ = 90°). A stress of 90 MPa is applied to nanomagnet Out for 3 ns. As shown in **Figure 10(a)–(d)**, the nanomagnet flips to "NULL" after the stress has been applied for 1.8 ns. Note that the magnetization of "NULL" state here does not exactly correspond to φ = 0. Rigorously, it makes a certain angle (φ = 7°) with the x axis. This is because the stress field component in the $-y$ direction and the field component of shape anisotropy in the $+y$ direction yield a stable equilibrium, so that the magnetization vector of magnet Out is stably deviated from the x axis. If φ < 10°, the field component of shape anisotropy energy in the $+y$ direction is much smaller than the stress field component, thus not affecting calculation result. Inputs "00," "01," "10," and "11" are read in at 2.9 ns. After the stress has been released for 0.9 ns (t = 3.9 ns), magnet Out will flip to a stable logic state. When the inputs are "01," "10," and "11," magnet Out is logic "1" (φ = 88°), whereas when the inputs are "00," the magnet Out is logic "0" (φ = −92°), successfully yielding OR logic. Note that the nanomagnet Out does not flip to the long axis (φ = 85° or φ = −95°) under the interaction of the ferromagnetic coupling of the input nanomagnets.

The input nanomagnets A and B only produce small fluctuations (~2°) in the plane and eventually return to the original logic state (φ = 90° or φ = −90°) under the interaction of the ferromagnetic coupling of nanomagnet Out. The angular variations of θ are similar in the four situations. Situation of inputs "10" is specially shown in **Figure 10(e)** and **(f)**. The polar angles (out-of-plane) of initial and final states of the three magnets are all θ = 90°. Magnets A and B produce smaller fluctuations (~2°) than magnet Out (~33°), as shown in **Figure 10(e)**.

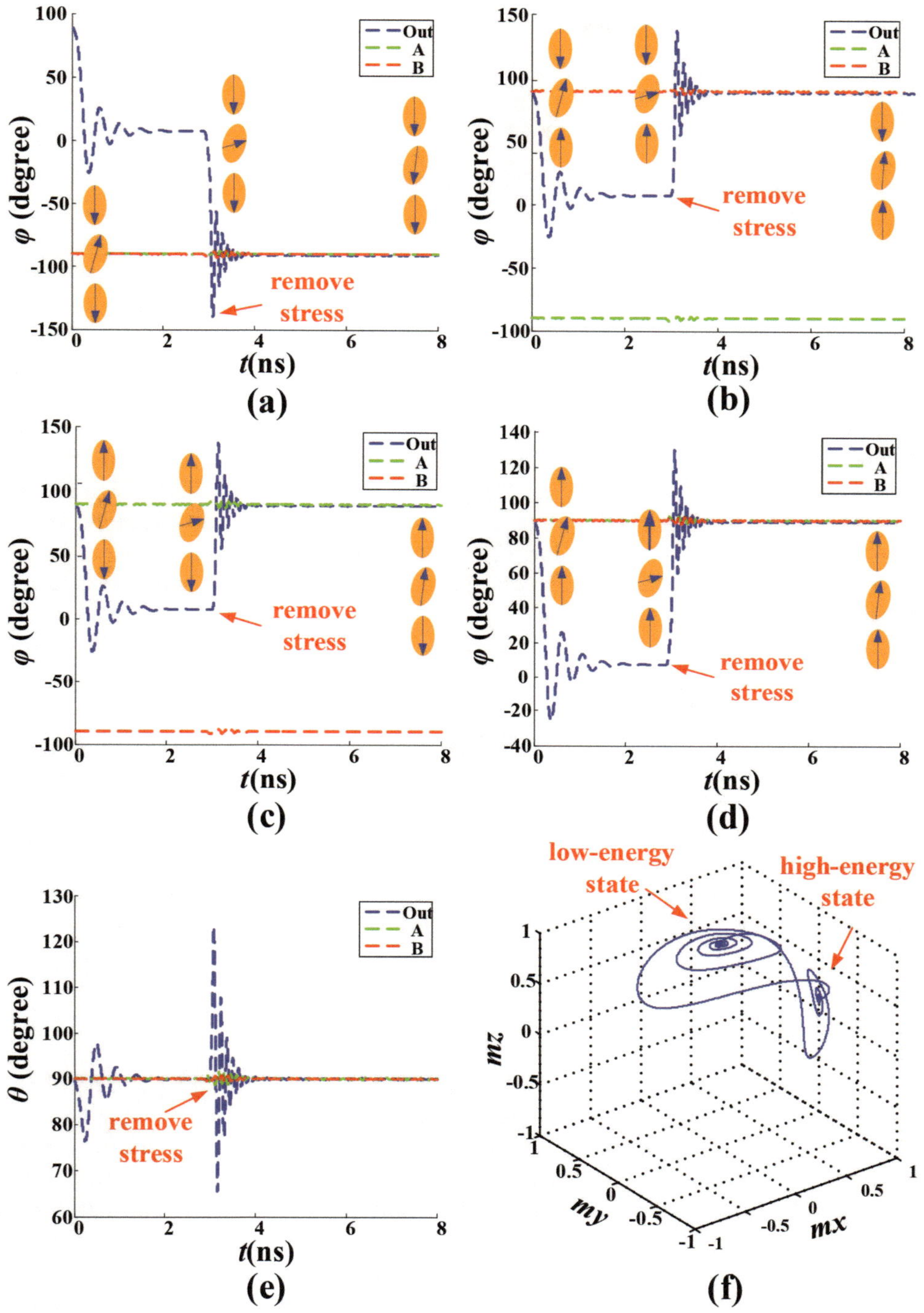

Figure 10.
Apply a stress of 90 MPa to magnet out for 3 ns. (a)–(d) show dynamic magnetization of the azimuth angle φ of (a) input "00," output "0"; (b) input "01," output "0"; (c) input "10," output "0"; and (d) input "11," output "1." when the input is "10," (e) and (f) show (e) dynamic magnetization of the polar angle θ and (f) magnetization track of the nanomagnet out.

The results confirm that magnets A and B will remain stable during the switching of magnet Out. The magnetization track of the magnet Out presents two obvious energy states, as can be seen from **Figure 10(f)**.

Figure 11 shows the simulation of our design of OR logic gate calculated by OOMMF using the data in **Table 1**. The other parameters are set as follows: space size = 800 nm × 200 nm × 20 nm, and mesh size = 5 nm × 5 nm × 5 nm.

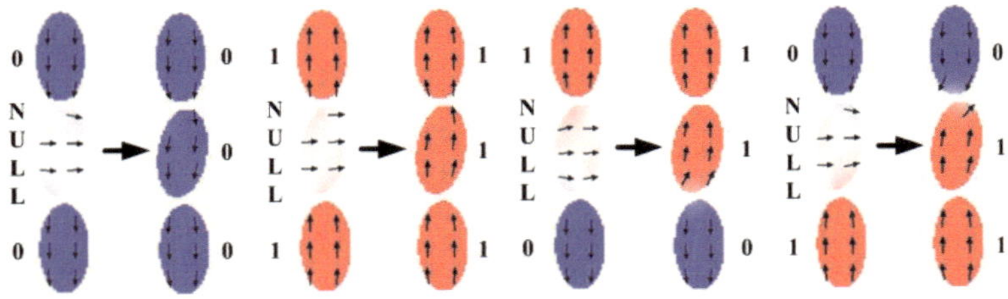

Figure 11.
Simulation results of OR logic gate by OOMMF.

The initial clock is pointing right and the inputs are "10," "01," "00," and "11." Only when the inputs are "00," the output becomes "0"; otherwise the output is "1," yielding OR logic as expected.

Unlike designs based on slanted nanomagnet, basic logic gates based on tilted nanomagnet have three advantages: (1) This tilted magnet design allows high aspect ratio (2:1) nanomagnets to be used; as a consequence of which, less C-shaped and eddy current clock errors will occur; (2) regular-shaped tilted nanomagnet reduces the requirements of fabrication process; and (3) the regular shape provides great convenience in numerical calculation.

3.3 Conclusion

In this section, a design of AND/OR logic gates is proposed based on tilted placement of nanomagnet. The mathematical model of the design is established, and the correctness is verified by the OOMMF software. This scheme can provide a more efficient and reliable basic logic unit for NML design. However, in the experimental preparation, there may be fabrication errors in tilting the placement of the nanomagnet. To reduce the process fabrication error, stress electrodes may be tilted so that the stress will also make an angle with the long axis of the nanomagnet.

4. Conclusions

In this chapter, the multiferroic heterojunction is introduced into the field of spintronics. By utilizing the inverse piezoelectric effect and the inverse magnetostrictive effect in the multiferroic heterojunction, the weak electric field can be used to accurately synchronize the storage and processing of the magnetic logic signal of the uniaxial nanomagnet. Multiferroic nanomagnets are considered to be a strong competitor for post-CMOS devices due to their natural nonvolatility, high radiation resistance, and ultra-low power consumption. In this chapter, the multiferroic nanomagnet device is taken as the research object, and the research on the two key problems of fast nanomagnet rapid reversal magnetization reversal and nanomagnetic logic gate is carried out. The research results have great innovation and application background.

Acknowledgements

This work was supported in part by the National Natural Science Foundation of China (Grant Nos. 61832007) and the National Key R&D Program of China (Grant No. 2018YFB1003304).

Author details

Jiahao Liu* and Liang Fang*
Institute for Quantum Information, State Key Laboratory of High Performance Computing, College of Computer, National University of Defense Technology, Changsha, China

*Address all correspondence to: rae20121220@163.com and lfang@nudt.edu.cn

References

[1] Ma J, Hu J, Li Z, et al. Recent progress in multiferroic magnetoelectric composites: From bulk to thin films. Advanced Materials. 2011;**42**(18): 1061-1087. DOI: 10.1002/adma.201190024

[2] Imre A, Csaba G, Ji L, et al. Majority logic gate for magnetic quantum-dot cellular automata. Science. 2006; **311**(5758):205-208. DOI: 10.1126/science.1120506

[3] Lee J, Dong IS, Park W. The universal magnetic tunnel junction logic gates representing 16 binary boolean logic operations. Journal of Applied Physics. 2015;**117**(17):17D717. DOI: 10.1063/1.4916806

[4] Liu JH, Yang XK, Cui HQ, et al. Modeling of 180° magnetization switching and clock sensitivity in a tilted multiferroic nanomagnet. Journal of Magnetism and Magnetic Materials. 2019;**474**:161. DOI: 10.1016/j.jmmm.2018.10.114

[5] Fidler J, Schrefl T. Micromagnetic modelling-the current state of the art. Journal of Physics D: Applied Physics. 2000;**33**:R135-R156. DOI: 10.1088/0022-3727/33/15/201

[6] Liu JH, Yang XK, Zhang ML, et al. Efficient dipole coupled nanomagnetic logic in stress induced elliptical nanomagnet array. IEEE Electron Device Letters. 2019;**40**(2):220-223. DOI: 10.1109/LED.2018.2889707

[7] Chikazumi S, Charap S. Physics of Magnetism. New York: Wiley; 1964

[8] Fashami MS, Roy K, Atulasimha J, et al. Erratum: Magnetization dynamics, Bennett clocking and associated energy dissipation in multiferroic logic. Nanotechnology. 2011;**22**:155201. DOI: 10.1088/0957-4484/22/30/309501

[9] Brown WF. Thermal fluctuations of a single-domain particle. Physics Review. 1963;**130**:1319-1320. DOI: 10.1103/PhysRev.130.1677

[10] Biswas AK, Bandyopadhyay S, Atulasimha J. Complete magnetization reversal in a magnetostrictive nanomagnet with voltage-generated stress: A reliable energy-efficient non-volatile magneto-elastic memory. Applied Physics Letters. 2014;**105**(7): 072408. DOI: 10.1063/1.4893617

[11] Biswas AK, Ahmad H, Atulasimha J, et al. Experimental demonstration of complete 180° reversal of magnetization in isolated co nanomagnets on a pmn-pt substrate with voltage generated strain. Nano Letters. 2016;**17**(6):3478. DOI: 10.1021/acs.nanolett.7b00439

[12] Roy K, Bandyopadhyay S, Atulasimha J. Binary switching in a symmetric potential landscape. Scientific Reports. 2013;**3**:3038. DOI: 10.1038/srep03038

[13] Zhu W, Xiao D, Liu Y, et al. Picosecond electric field pulse induced coherent magnetic switching in mgo/fept/pt(001)-based tunnel junctions: A multiscale study. Scientific Reports. 2014;**4**(7):4117. DOI: 10.1038/srep041171

[14] Barangi M, Mazumder P. Straintronics: A leap toward ultimate energy efficiency of magnetic random access memories. IEEE Nanotechnology Magzine. 2015;**9**(3):15-24. DOI: 10.1109/MNANO.2015.2441106

[15] Li X, Carka D, Liang C, et al. Strain-mediated 180° perpendicular magnetization switching of a single domain multiferroic structure. Journal of Applied Physics. 2015;**118**(1): 232905-232943. DOI: 10.1063/1.4923350

[16] Peng RC, Hu JM, Momeni K, et al. Fast 180° magnetization switching in a strain-mediated multiferroic heterostructure driven by a voltage. Scientific Reports. 2016;**6**:27561. DOI: 10.1038/srep27561

[17] Jin TL, Hao L, Cao JW, et al. Electric field control of anisotropy and magnetization switching in CoFe and CoNi thin films for magnetoelectric memory devices. Applied Physics Express. 2014;7:043002. DOI: 10.7567/APEX.7.043002

[18] Quintero SMM, Martelli C, Braga AMB, et al. Magnetic field measurements based on terfenol coated photonic crystal fibers. Sensors. 2011; **11**(12):11103-11111. DOI: 10.3390/s111211103

[19] Roy K, Bandyopadhyay S, Atulasimha J. Switching dynamics of a magnetostrictive single-domain nanomagnet subjected to stress. Physics Review B. 2011;**83**:224412. DOI: 10.1103/PhysRevB.83.224412

[20] Donahue MJ, Porter DG. 1999. OOMMF User's Guide, Version 1.0. Interagency Report, NISTIR 6367. Available from: http://www.math.nist.gov/oommf

[21] AI-Rashid MM, Bandyopadhyay S, Atulasimha J. Dynamic error in strain-induced magnetization reversal of nanomagnets due to incoherent switching and formation of metastable states: A size-dependent study. IEEE Transactions on Electron Devices. 2016; **63**(8):3307-3313. DOI: 10.1109/TED.2016.2582142

[22] Liu JH, Yang XK, Cui HQ, et al. Voltage pulse-induced fast and repeated switching in a uniaxial nanomagnet at room temperature. Journal of Magnetism and Magnetic Materials. 2019;**491**:165607. DOI: 10.1016/j.jmmm.2019.165607

[23] Gypens P, Leliaert J, Waeyenberge B. Balanced magnetic logic gates in a kagome spin ice. Physics Review Applied B. 2018;**9**:034004. DOI: 10.1103/PhysRevApplied.9.034004

[24] Roy K. Ultra-low-energy non-volatile straintronic computing using single multiferroic composites. Applied Physics Letters. 2013;**103**:173110. DOI: 10.1063/1.4826688

Chapter 3

Perpendicular Magnetic Insulator Films for Spintronics

Laith Alahmed and Peng Li

Abstract

The recent progress in spintronics opens up new directions for novel device concepts and fundamental understandings. This is possible because of magnetic insulators (MIs), which have paved the way toward pure spin current-based spintronics. MIs with perpendicular anisotropy expand the horizon further, enabling new functionalities such as low-power spin-orbit torque switching, high-speed domain-wall motion, high-frequency spin-orbit torque oscillation, etc. In this chapter, we review recent progress in spintronic experiments using barium hexagonal ferrite $BaFe_{12}O_{19}$—a magnetic insulator with perpendicular anisotropy. These results lay the foundation for using MIs with perpendicular anisotropy as a medium to develop new energy-efficient pure spin current-based electronics.

Keywords: magnetic insulator, perpendicular anisotropy, spin current, spin-orbit torque, spintronics

1. Introduction

Spintronics, also known as spin electronics, is a newly emerging field of research that focuses on the spin degree of freedom of electrons rather than their charge. Charge current is a flow of electrons from one point to another under the influence of an electric field. In spintronics, spin current can propagate within the material. A pure spin current can be generated through effects such as the spin Hall effect (SHE), spin pumping, spin-wave propagation, etc. The pure spin currents consume much less energy than charge currents. This is because of the absence of charge flow that eliminates the power consumption needed for the electric field required to drive charge flow [1–3].

In spintronics, magnetization switching is of both fundamental interest and technological significance. One way to switch the magnetization of a ferromagnetic film is through the spin filtering effect. In this case, a spin-polarized electrical current will be generated. As the polarized electrons flow through the ferromagnetic film, they transfer angular momentum to the film and produce a spin-transfer torque to switch the film. This torque is called spin-transfer torque (STT). Magnetic random-access memory based on STT has already been commercialized in recent years.

The above-mentioned spin-torque switching, however, has a limit. The angular momentum transferred per unit charge in the applied current usually cannot exceed a quantum of spin ($\hbar/2$). Recent work demonstrates that one can exceed this limit by the use of spin-orbit torque (SOT). The demonstration generally takes a nonmagnetic heavy metal (HM)/ferromagnetic metal (FM) bilayered structure and makes use of the spin-orbit coupling-produced SHE in the HM film to convert an in-plane charge current to a pure spin current that flows across the HM thickness. This produces spin accumulation at the HM/FM interface and therefore exerts a SOT on FM. In this case, each electron in the applied current can undergo multiple spin-flip scattering at the interface, therefore enabling more efficient switching than in the conventional spin-transfer torque case.

The ferromagnetic films used in most of the SOT studies were all conductive. A direct consequence is the severe shunting current in the ferromagnet layer, which not only limits the switching efficiency but also causes parasitic effects. For example, previous works have shown that interfacing a TI with a conductive FM film can result in a significant modification or even complete suppression of the topological surface states (TSSs) in the TI layer. In a TI/FM heterostructure, the TSSs may have been largely spoiled by the FM electrons. This means that many large spin-orbit torques observed in TI/FM structures may not be due to TSS. In this context, the use of MIs in an HM/MI heterostructure can effectively avoid the shunting current. Moreover, the TSSs in a TI/MI structure can be preserved except for the opening of a small gap at the Dirac point when strong coupling exists at the interface. This will enable the magnetization switching due to *bona fide* TSSs.

Magnetic insulators include a large class of materials, including spinels, garnets, and ferrites. They have a general chemical formula of $M(Fe_xO_y)$, where M is representing non-iron metallic elements. MIs have several advantages over magnetic metals for SOT device applications. First, in a heavy metal/MI heterostructure, the charge current only flows in the HM layer but not in the MI layer. In contrast, in an HM/magnetic metal structure, the charge current also flows in the FM, resulting in certain parasitic effects. When the HM layer is replaced by a topological insulator with high resistivity, the advantage of zero shunting currents in the MI film becomes particularly important. Moreover, interfacing a topological insulator (TI) with a conductive FM can result in a significant modification or even complete suppression of the topological surface states (TSSs) in the TI layer. The use of a magnetic insulator can effectively avoid the shunting current; TSSs in a TI/magnetic insulator (MI) structure can also be well preserved.

In the ferrite family, hexagonal ferrites have strong magnetocrystalline anisotropy. For example, M-type barium ferrite ($BaFe_{12}O_{19}$, noted as BaM) has an anisotropy field of 17 kOe. The perpendicular anisotropy in MI films originates from bulk intrinsic anisotropy rather than interfacial anisotropy [4]. This means that, when being used for actual devices, the BaM film has no constrains on the thickness. This is in strong contrast with the ferromagnetic metal counterpart (e.g., CoFeB/MgO) that often has to be very thin to realize interfacial perpendicular anisotropy. In addition, the magnetic damping is usually significantly lower in MIs than in FMs. For example, the intrinsic Gilbert damping constant in BaM materials is 7×10^{-4}, which is at least 10 times smaller than the value in permalloy [5]. This advantage is significant for spin-torque oscillator applications, where the current threshold for self-oscillations decreases with the damping, as well as for logic device applications that require low-damping, insulating spin channels.

This chapter reviews the main advances made in spintronic experiments with BaM over the past several years. Section 2 gives a brief introduction to BaM and discusses its crystalline structure, magnetic properties, and thin film growth techniques. This section serves to provide a background for the discussions in the following sections. Section 3 reviews the advances of spintronic experiments with BaM. Section 3.1 provides an overview of the related spintronic experiments. Section 3.2 discusses the generation of pure spin currents through the spin Seebeck effect and photo-spin-voltaic effect in the Pt/BaM structure. Section 3.3 discusses the spin-orbit torque-assisted switching in BaM. Section 3.4 discusses the use of topological insulator/BaM heterostructure for magnetization switching. Finally, Section 3.5 provides an outlook in the field of BaM materials and devices.

2. Properties of barium ferrite thin films

2.1 Atomic structure of BaM thin films

BaM is a hexagonal ferrite, which consists of close-packed layers of oxygen ions. **Figure 1** shows a unit cell of BaM. The Ba^{2+} ion is large, as is the O^{2-} ion, and the barium always replaces oxygen somewhere in the oxygen lattice. The close-packed layers form six fundamental blocks, namely, S, S^*, R, R^*, T, and T^* [5–7]. The S block consists of close-packed oxygen layers stacking in an ABCABC ... sequence. It has a cubic spinel arrangement with the < 1 1 1 > axis along the vertical direction. There are two units of Fe_3O_4 without any barium ions in each S block. The R block comprises close-packed oxygen layers stacking in an ABAB ... sequence. It has a hexagonal closest packed structure along the vertical axis. Each R block has a unit formula of $BaFe_6O_{11}$. The T block is made of four oxygen layers, with a barium ion replacing an oxygen ion in the middle two layers, which gives a unit formula of $Ba_2Fe_8O_{14}$. The S^*, R^*, and T^* blocks are 180° rotations around the c-axis from the S, R, and T blocks. BaM is built from the stacking of S, R, S^*, and R^* blocks.

Trivalent Fe^{3+} ions occupy tetrahedral and octahedral sites as well as one trigonal bipyramidal site. Different sites account for different spin orientations and Bohr magnetons (μ_B). For example, a tetrahedral site contributes $2\mu_B$, while an octahedral site contributes $4\mu_B$ with opposite spin orientations in the S block. In the end, S, S^*, R, and R^* blocks contribute $2\mu_B$ each, leading to a moment of $40\mu_B$ for each unit cell. This gives a saturation magnetization of ~4700 G in bulk BaM. BaM has a strong

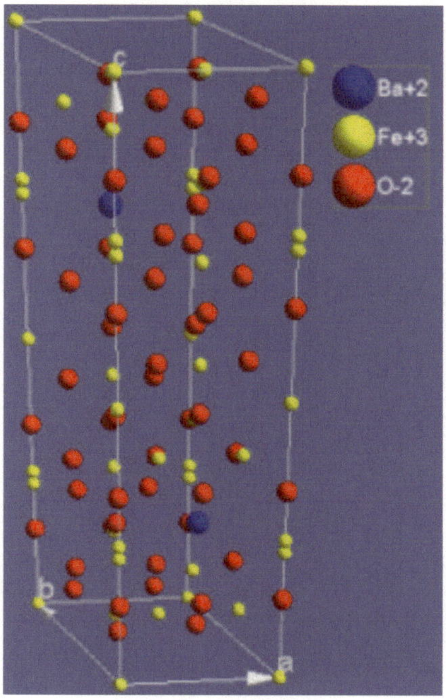

Figure 1.
Crystalline structure of M-type barium ferrite. Blue ball, Ba^{2+}. Yellow ball, Fe^{3+}. Red ball, O^{2-}.

anisotropy field of 17 kOe, which is along the *c* axis. This comes from the trigonal bipyramidal site Fe^{3+} ions, as well as breaking crystal symmetry in the R/R* blocks. This is the most distinguished property of BaM, because the perpendicular anisotropy field originates from bulk intrinsic anisotropy. BaM has a large *c* constant of 23.2 Å and an *a* constant of 5.89 Å. The *x*-ray density is about 5.29 g/cm^3. The Curie temperature of bulk BaM is 725 K, which is much higher than the room temperature. The exchange constant is 6.4×10^{-7} erg/cm [7].

2.2 Growth techniques

A variety of techniques are used to grow BaM thin films, including pulsed laser deposition (PLD) [8–10], alternating target laser ablation deposition (ARLAD) [11, 12], molecular beam epitaxy (MBE) [13], liquid phase epitaxy (LPE) [14, 15], magnetron sputtering [16, 17], and so on. Guo et al. at Boston Applied Technologies proposed a chemical solution deposition process to deposit BaM. Song and his colleagues succeeded in the PLD growth of BaM thin films that showed an FMR linewidth as narrow as single-crystal BaM bulks. However, these films showed a remanent magnetization much smaller than the saturation magnetization [9]. This problem was improved in the later experiments when tuning the deposition conditions [18]. **Figure 2** shows the PLD parameters which decide the thin film quality. **Figure 2b** shows that *c*-axis out-of-plane BaM grains can be grown on (0001) Al_2O_3 substrates; *c*-axis in-plane BaM grains can be grown on (11 −20) Al_2O_3 substrates. A typical procedure is as follows: the oxygen pressure is set at 300 mTorr, and the substrate is heated to 800°C. The substrate-to-target separation is fixed at 4 cm, and the energy fluence of the laser beam is set to 0.7 J/cm^2. The laser pulse repetition rate is increased from 1 to 5 pulse(s) per second in five equal steps over the first 5 min and is then set to 10 pulses/s for the remaining deposition. After the

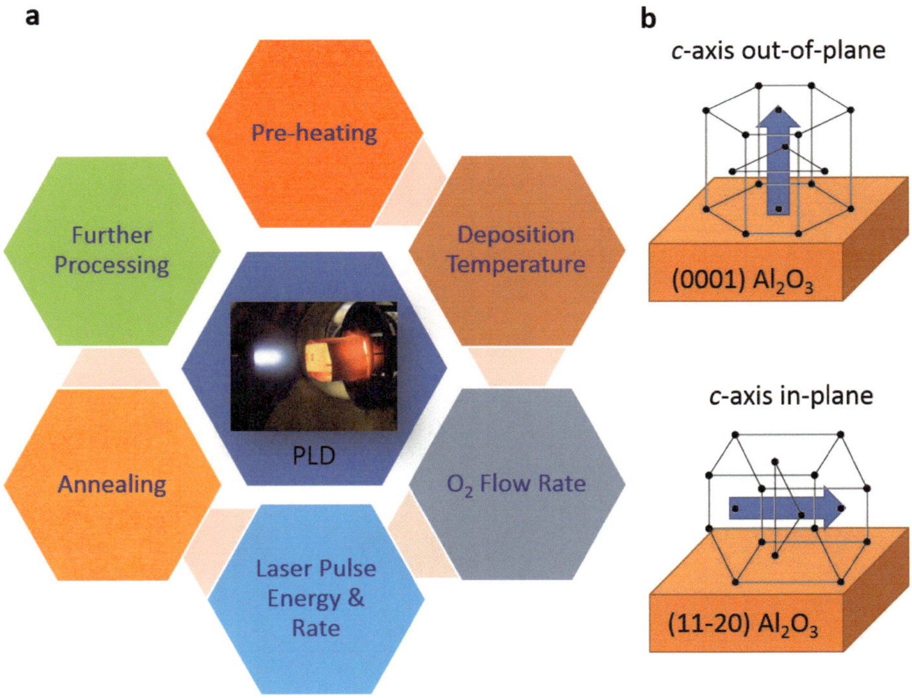

Figure 2.
Growth condition in pulsed laser deposition of BaM thin films. (a) Parameters controlling the BaM thin film quality. (b) Different Al_2O_3 substrate types for growing BaM with different c*-axis orientations.*

deposition, the substrate is cooled down at a rate of 2°C/min in 400 Torr oxygen. The sample is then annealed at 850°C for 4 h in a standalone tube furnace, with a heating rate of 10°C/min and a cooling rate of 2°C/min.

2.3 BaM thin film grown on (0001) *c*-plane Al_2O_3 substrate

In microwave device applications, BaM films usually have a thickness of several microns. For spintronic devices, the thickness is reduced to tens of nanometers. **Figure 3** shows the structure and magnetic properties of nanometer-thick BaM thin films grown on a *c*-axis Al_2O_3 substrate. The atomic force microscopy (AFM) image in **Figure 3a** shows a uniform and smooth surface, and the analysis of the AFM data yielded an RMS surface roughness of 0.19 ± 0.03 nm. These results, together with other AFM data not shown, indicate that the BaM film has a reasonably good surface, which is critical for the realization of high-quality BaM thin films. The roughness value here is an average over the measurements of nine different 1×1 μm areas, and the uncertainty is the corresponding standard deviation.

Figure 3b shows a $2\theta/\omega$ *x*-ray diffraction (XRD) scan, with the XRD intensity on a log scale. The *x*-ray θ rotation gave a scattered beam that matched the specular reflection from the surface. The detected (001) diffraction peaks all come from *c*-plane scattering of the BaM film. The (006) sapphire substrate peak was also detected. The hysteresis loops in **Figure 3c** were measured by a vibrating sample magnetometer with different field orientations, as indicated. The loops clearly show that the BaM film has perpendicular anisotropy, which confirms the *c*-axis orientation of the film. Analysis of the hysteresis data yielded an effective perpendicular anisotropy field around H_{ani} = 20 kOe, which is larger than the bulk value (17 kOe). The normalized saturation magnetization $4\pi M_s$ = 4.16 kG, which is lower than the bulk value of BaM (4.70 kG). **Figure 3d** presents a ferromagnetic resonance (FMR)

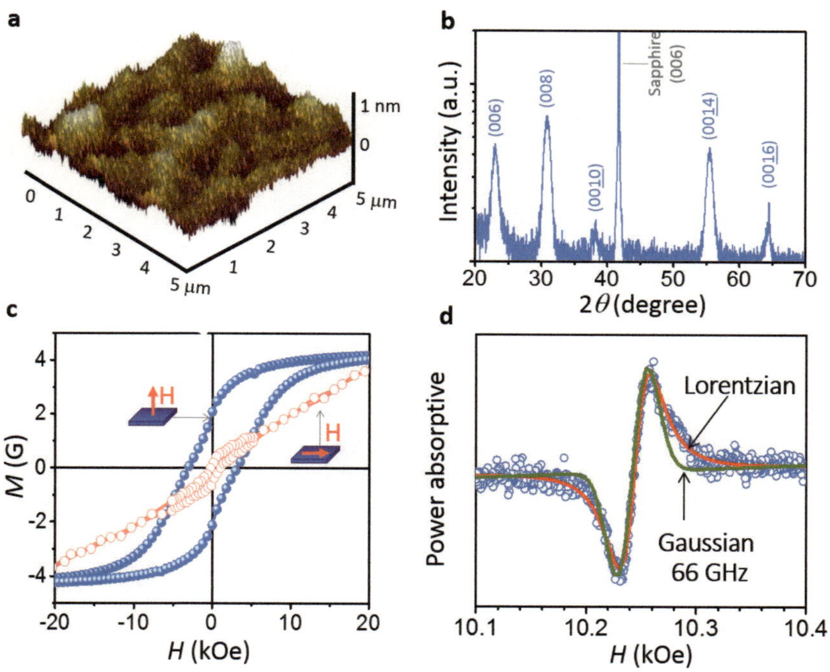

Figure 3.
Structure and magnetic properties of BaM thin films. (a) Atomic force microscope of 5 nm BaM thin film. (b) x-ray diffraction of 5 nm BaM thin film. (c) Hysteresis loops of 5 nm BaM thin film. Blue circles, H along out-of-plane direction. Red circles, H along in-plane direction. (d) Ferromagnetic resonance of 20 nm BaM thin film with H along out-of-plane direction. a, b, and c are adapted from [10].

curve obtained with a 20-nm-thick BaM film at ω = 66 GHz. Because of the strong perpendicular anistropy field, the ferromagnetic resonance of BaM film appears between 50 GHz and 75 GHz. In the graph, the blue circles show an FMR profile measured at 66 GHz. The Lorentzian function (red curve) fits the data points better than the Gaussian fit shown as the olive curve, indicating that the film has a uniform quality. The fitting yielded a peak-to-peak linewidth ΔH = 26.59 $\pm$ 0.60 Oe and an FMR resonance field H_{res} = 10.24 kOe. Similar measurements can be carried out to FMR profiles at a variety of frequencies, and a Kittel equation can fit the curve with a magnetic field applied out-of-plane:

$$\omega = 2\pi|\gamma|(H_{\text{res}} + H_{\text{ani}} - 4\pi M_{\text{s}}) \tag{1}$$

Such fitting yielded a gyromagnetic ratio γ = 2.80 $\pm$ 0.01 MHz/Oe and H_{ani} = 19.12 $\pm$ 0.04 kOe. The linewidth ΔH vs. frequency ω data can be fitted with the following equation:

$$\Delta H = \frac{2\alpha}{\sqrt{3}|\gamma|}\frac{\omega}{2\pi} + \Delta H_0 \tag{2}$$

where α is the damping constant and H_0 is the inhomogeneity line broadening, which is a parameter associated with α describing the damping of the material. The FMR measurements yielded a damping constant α = $(9.7 \pm 1.1)\times 10^{-4}$.

2.4 BaM thin film grown on (1 1 −2 0) *a*-plane Al_2O_3 substrate

Figure 4 shows the structural and magnetic properties of a representative BaM film that is grown on an (1 1 −2 0) *a*-plane Al_2O_3 substrate. The film has a thickness

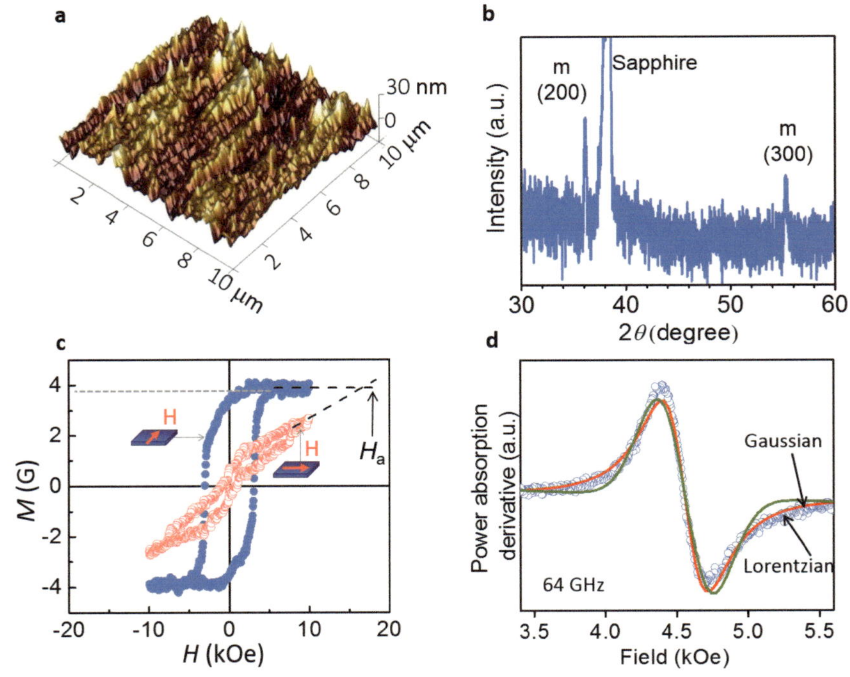

Figure 4.
Structure and magnetic properties of 1.2 μm BaM thin films with c-axis in plane. (a) AFM image. (b) XRD spectrum. (c) Hysteresis loops of the BaM thin film. Blue circles, H along the in-plane easy axis direction. Red circles, H along in-plane hard axis direction. (d) Ferromagnetic resonance spectrum with Gaussian and Lorentzian fittings. a, b, and c are adapted from [18].

of 1.2 μm. Thus, the AFM data show a relatively rough surface with an RMS surface roughness of 14.3 ± 1.6 nm. **Figure 4b** presents an XRD spectrum. The spectrum consists of a strong peak from the sapphire substrate and the two other peaks for the m-planes of the BaM film, indicating the in-plane orientation of the c-axis. **Figure 4c** presents the two hysteresis loops of the film measured by a vibrating sample magnetometer. One of the loops was measured with the magnetic field applied along the c-axis, while the other was measured with the field also in the film plane but perpendicular to the c-axis. The dashed lines indicate the extrapolations used to determine the effective anisotropy field H_{ani}. The dotted line indicates the determination of the saturation induction $4\pi M_s$. These data show that the film has a well-defined in-plane uniaxial anisotropy with the easy axis along the c-axis. From the hysteresis loops, it can be concluded that H_{ani} = 16.5 kOe and $4\pi M_s$ = 3.87 kG. Moreover, the film has a large remanent magnetization-to-saturation magnetization ratio of about 0.89 along the c-axis. This large ratio results from the strong in-plane uniaxial anisotropy. This means that the film is self-biased to a large degree at zero fields, which allows for the use of BaM in self-bias spintronic experiments. **Figure 4d** shows the FMR spectrum of the c-axis in-plane BaM film shown by the blue, open circles measured at 64 GHz. The curve is fitted to the derivatives of a trial Lorentzian function and a Gaussian function. Similar to the FMR curve in Section 2.3, it fits better with the Lorentzian function, indicating a uniform film. The Lorentzian fitting yields H_{res} = 4.55 kOe and ΔH = 318 Oe. The dependence of H_{ani} and the microwave frequency f can be fitted by a Kittel equation for field applied in-plane:

$$\omega = 2\pi|\gamma|\sqrt{(H_{\text{res}} + H_{\text{ani}})(H_{\text{res}} + H_{\text{ani}} + 4\pi M_s)} \tag{3}$$

3. Spintronic applications with magnetic insulators

3.1 Introduction to spintronics

In the following sections, we introduce recent spintronic experiments using MIs with strong anisotropy fields. Devices that incorporate the unique properties of MIs are an excellent potential solution for the power consumption and heat dissipation problems of conventional electronics, as they would consume much less energy and generate significantly less heat. We introduce the use of different techniques in generating pure spin currents, using bilayer heterostructures of a normal metal (NM)/ferromagnetic material. There are a variety of normal metal choices such as platinum (Pt) and Gold (Au). Both have been explored and tested in spintronics related studies and experiments [18–33].

In the first two sections, we will explore the generation of pure spin currents using the spin Seebeck effect (SSE) and the photo-spin-voltaic effect (PSVE). Both techniques take advantage of a NM coupled with a MI. In SSE, a temperature gradient in the MI is the main factor that induces the MI to inject pure spin currents into the NM layer. In PSVE however, the light of certain wavelengths reaching the atomic layers of the NM, exciting the NM electrons near the NM/MI interface, is what generates the pure spin currents. SSE and PSVE Experimentation results will also be explored and discussed. Then, in the last two sections, we will demonstrate how pure spin currents can be used practically to enhance magnetic switching in MIs in a significant and meaningful way. NM/MI bilayers will not be the only type of heterostructure discussed here, we will also explore topological insulator/MI structures and demonstrate the significance of topological insulators in spintronics.

3.2 Generation of pure spin currents through SSE using NM/BaM structures and photo-spin-voltaic effect in Pt/BaM structure

3.2.1 The spin Seebeck effect

The traditional Seebeck effect, first discovered by Thomas Seebeck in 1821 [34], refers to the generation of electric potential in a conductor when a temperature gradient is applied to it. The electric potential is caused by charge carriers within the conductor moving from the hot region to the cold region. A thermocouple consists of two dissimilar conductors that are joined to form a junction; when a heat gradient is applied across the thermocouple (see **Figure 5a**), a voltage difference can be observed across them. The sign of the voltage flips when the direction of the temperature gradient is flipped. The traditional Seebeck effect is the basic principle behind most thermoelectric generators.

The spintronic equivalent of the traditional Seebeck effect, called the spin Seebeck effect, was first discovered in 2008 [19, 28]. SSE is a phenomenon that can be observed in ferromagnetic and ferrimagnetic materials when a heat gradient is applied to them [19, 28, 35]. The heat gradient induces a spin voltage in the ferromagnet that can be used to inject pure spin currents into a conductor attached to the ferromagnet. Here, spin voltage is a potential for the spin of electrons, rather than their charge, to drive spin current [19, 36–38]. Previously mentioned bilayer heterostructures of normal metal/magnetic material have been used to study the SSE in two different configurations: transverse and longitudinal [19, 39]. In the transverse configuration, the generated spin current is perpendicular to the temperature gradient [28]. The generated spin current in the longitudinal configuration is parallel to the temperature gradient [19] (see **Figure 5b**). The longitudinal configuration has been the dominant choice for SSE research, owing to its simplicity [19]. Magnetic insulators (such as YIG, BaM, etc.) offer an ideal platform for observing the longitudinal spin Seebeck effect (LSSE) [19, 40]. In a conductive ferromagnet, the longitudinal configuration can give rise to a large anomalous Nernst effect (ANE)-induced voltage, which makes it difficult to distinguish between ANE and SSE [19, 33, 41, 42].

If SSE generates pure spin currents, then an important question would be how do we measure them? The absence of charge flow makes it impossible to use conventional methods to measure the spin currents. One way to measure LSSE-generated spin current is to first convert it into a charge current that can then be measured by conventional means. In this context, the choice of the normal metal in

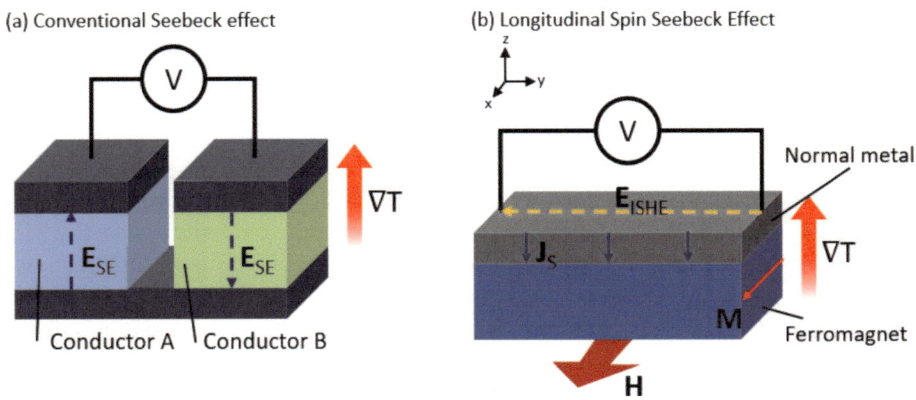

Figure 5.
Schematic illustrations of (a) the conventional Seebeck effect and (b) longitudinal spin Seebeck effect.

the bilayer heterostructure becomes very important. Heavy metals, such as Pt and Au, have strong spin-orbit coupling [43, 44], offering an effective mechanism to convert a transverse spin current into a longitudinal charge current through inverse spin Hall effect (ISHE) [43, 45–47]. The ISHE charge current across the heavy metal surface creates an electric field E_{ISHE} that can be measured with a voltmeter. The magnitude and sign of E_{ISHE} depend on internal and external factors. In **Figure 5b**, an external magnetic field H is applied in the x direction. The magnetization M of the MI layer is aligned to the x direction as well. The temperature gradient is applied across the z direction, generating a spin voltage in the MI layer, injecting spin current across the interface and into the normal metal layer parallel to the temperature gradient. The polarity of electron spins in the normal metal layer is influenced by M from the MI layer. The ISHE field, E_{ISHE}, is measured across the y direction. E_{ISHE} is proportional to the cross product of the spatial direction of the generated spin current J_s and the polarity vector of electron spins in the normal metal layer. The following equation explains the relationship between E_{ISHE}, J_s, and σ [19]:

$$E_{\text{ISHE}} \propto J_s \times \sigma \tag{4}$$

In summary, the voltage measured across the normal metal surface is strongest when M is perpendicular to both the heat gradient and E_{ISHE}; the voltage will flip its sign if M is flipped by flipping the external magnetic field H; the voltage measured will be zero when M is parallel to E_{ISHE}.

This discussion sheds light on the importance of the existence of an external magnetic field H to enable E_{ISHE} when using soft magnetic insulators such as spinels and garnets. A strong enough H is necessary to saturate the magnetization of such insulators, as well as to control the direction of the magnetization. Indeed, SSE cannot be observed in samples incorporating spinels or garnets with a temperature gradient alone. Due to their low remnant magnetization, an appropriate external magnetic field is required to saturate them.

An exception to the external magnetic field requirement is made when using BaM thin films due to their strong uniaxial anisotropy [18]. In the absence of an external magnetic field, the magnetization of BaM films, caused by the spins of unpaired electrons, tend to favor one axis, called the easy axis, over any other axis. Thus, most electron spins within the BaM film tend to align themselves with the easy axis, randomly up or down, in the absence of an external magnetic field. Therefore, BaM films have uniaxial anisotropy. The uniaxial field of BaM was found to be around 16.5 kOe [9, 18]. Applying a magnetic field of this value or higher along the easy axis of the film causes all the electron spins to align themselves in the direction of the magnetic field, removing the magnetic field then will leave a large remnant magnetization within the BaM film owing to its uniaxial anisotropy. Namely, the film becomes self-biased and does not require an external field to magnetize it.

An LSSE experiment and its results using a Pt/BaM heterostructure [18] will be discussed next. In this experiment the sample consisted of a micron-thick BaM layer, topped with a 2.5-nm-thick Pt layer. The BaM layer was grown on a 0.5 mm sapphire substrate. The easy axis of the BaM film was in the plane of the film.

Figure 6 shows the experiment setup and results. **Figure 6a** shows a schematic diagram of the experimental setup that was used to test LSSE within the sample. The sample was put on an aluminum plate to act as a heat sink. An incandescent light bulb was placed directly on top of the sample, acting as the heat source. The easy axis of the BaM layer was along the y-axis, and the voltage was measured along the x-axis. All measurements were performed without an external magnetic field. However, a magnetic field of 10 kOe was used prior to the experiment to set the magnetization M of the BaM film in the positive (or negative) y direction.

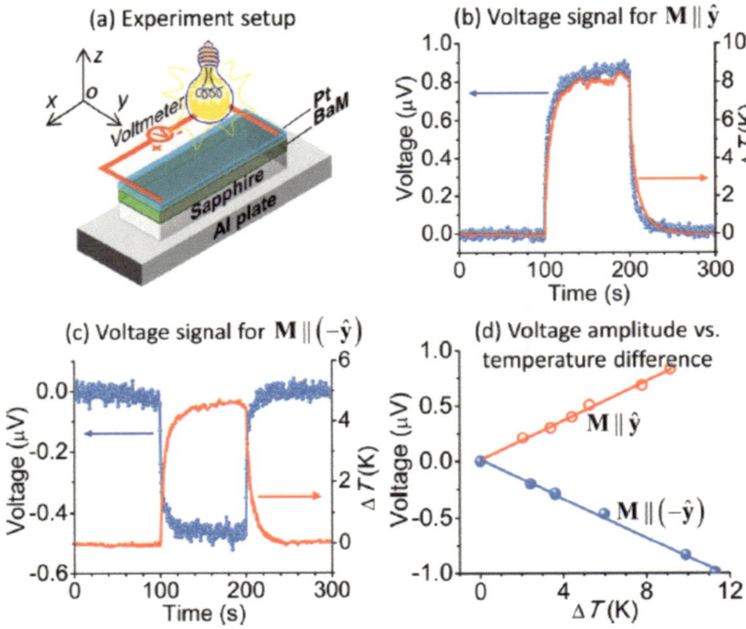

Figure 6.
Light-induced generation of spin currents. (a) The experimental setup. (b) and (c) Respective voltage signals measured for M||y and M||(−y), in response to the light that was turned on at 100s then turned off at 200s. The graphs also show the responses of the temperature difference (ΔT) between the top and bottom of the Pt/BaM/Saphire structure. (d) Voltage amplitude as a function of ΔT. Source: [18], p. 3.

The heat from the light bulb, along with the aluminum plate acting as a heat sink, created the temperature gradient across the BaM film thickness; the difference in temperature between the bottom surface and top surface of BaM, ΔT, was measured using two thermocouples connected to them. When the light is turned on, SSE occurs, the heat gradient induces a spin voltage in the BaM film that injects spin currents across the interface and into the Pt layer. Due to ISHE, the spin current is converted to a charge current across the Pt surface creating a voltage. The voltage was measured by connecting a nanovoltmeter to the opposite ends of the Pt surface across the x-axis.

Figures 6b and **c** demonstrate the relationship between the difference in temperatures ΔT and the generated voltage, with time, for $M||y$ and $M||-y$, respectively. ΔT was changed by changing the height of the light bulb. These results show that the SSE-generated spin current, and therefore the observed voltage, changes in exactly the same manner that ΔT changes with. This result is expected as the difference in temperature is what causes the spin voltage in the BaM film, which ultimately gives us the voltage reading across the Pt layer.

Figure 6d shows an important property of SSE, namely, the sign of the generated voltage flips when the direction of the BaM magnetization is flipped. The graph shows the relationship between ΔT and measured voltage for $M||y$ (red) and $M||-y$ (blue). The result again proves that the voltage change with respect to ΔT is identical (mirrored when $M||-y$ due to the sign flip), in both cases.

Control measurements were performed and are shown in **Figure 7**. Changing the lateral position of the light bulb did not have any noticeable effect on the measured voltage. This is to be expected, as the temperature gradient depends on the height of the light bulb, rather than its lateral position. This is demonstrated in **Figure 7a**, where the light position was changed to six different lateral positions. The figure shows that, other than jumps from electrical disturbance caused by the position change, the measured voltage remained largely unchanged.

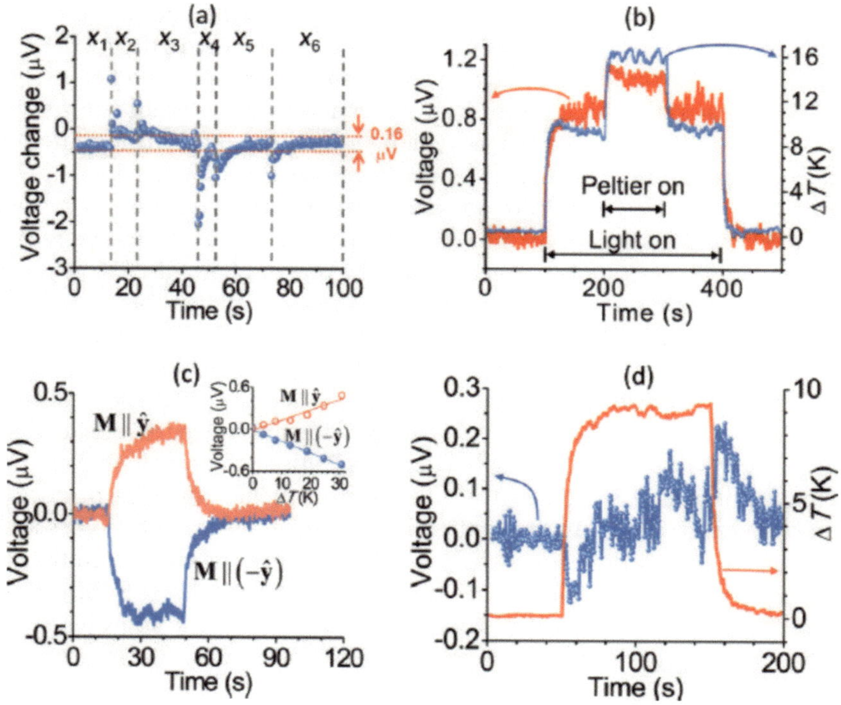

Figure 7.
Control measurements. (a) Voltage changes caused by moving the bulb along the x-axis. (b) Voltage and ΔT signals obtained when both a bulb and a Peltier cooler were used to control the temperature. The data in (a) and (b) were obtained with the same sample as ***Figure 6****. (c) SSE in a Pt(2.5 nm)/BaM(0.4 μm)/sapphire (0.5 mm) sample. (d) Voltage and ΔT signals obtained with a Cu(9 nm)/BaM(1.2 μm)/sapphire (0.5 mm) sample. Source: [18], p. 3.*

Using a Peltier cooler as an added source for the temperature gradient in addition to the light bulb also did not have a noticeable change in the relationship between the measured voltage and ΔT. This is shown in **Figure 7b**, where a light source was used to create ΔT, and a Peltier cooler was turned on under the sample midway through. **Figure 7c** shows the voltage change with time for a similar sample with a BaM layer thickness of 0.4 μm, and only a Peltier cooler was used to create the temperature gradient. Both **Figure 7b** and **c** show the same result: the measured voltage is directly related to ΔT, regardless of the method used to achieve ΔT.

The importance of using a metal with strong spin-orbit coupling is demonstrated through **Figure 7d**, where Cu, which has very weak spin-orbit coupling, and therefore very weak ISHE, was used in a Cu (9 nm)/BaM (1.2 μm)/sapphire (0.5 mm) sample. The figure shows a behavior that is different from the Pt/BaM samples, indicating the absence of SSE in this sample. A likely source for the signal shown in **Figure 7d** is the conventional Seebeck effect, caused by a temperature gradient across the sample's length. (All figures, experimentation setup and results were taken from [18] with appropriate permissions).

3.2.2 Photo-spin-voltaic effect

A closely related but fundamentally different effect to SSE is the photo-spin-voltaic effect (PSVE). PSVE happens in NM/MI heterostructures; it generates pure spin currents across the NM thickness that can be measured through ISHE. Light can generate spin voltage and drive spin currents through PSVE. While the spin voltage is generated in the MI layer in the SSE case, the spin voltage in PSVE is generated in the atomic layers of the NM that are close to the interface due to magnetic proximity effect [48]. When light of a certain wavelength hits the sample, photons excite electrons in the Pt layer, causing them to move to higher energy bands. The efficiency of this photon-driven excitation varies because of the spin orientation. The difference in efficiency, along with different diffusion rates of excited electrons and holes, generates the spin voltage through PSVE [48].

Figure 8 shows PSVE in a Pt/MI structure. An important question arises due to the extremely similar setup of both LSSE and PSVE: how can we determine the source of the ISHE generated voltage? It could be due to LSSE, or PSVE, or both. Fortunately, research in this area determined several distinguishable factors that make it possible to disentangle LSSE from PSVE. The most important factor is the wavelength of the light used to excite the sample. Experimental results determined that PSVE can only be observed when the wavelength of the light used falls in the range 1600–2000 nm [48]. Using a light source with a wavelength outside that range or a heat source other than light, such as a Peltier cooler, will only give us LSSE in our sample and no PSVE [49]. Other factors include the type of materials and device geometries used in the studies. For example, different MI types and thicknesses give widely different signals in LSSE. A recent work showed that the main contribution in the voltage comes from LSSE rather than PSVE [50]. However, experiments have shown that using a light source with the appropriate wavelength gives extremely similar results in Pt that is coupled with MI of varying types and thicknesses [48].

Figure 9 shows the results of PSVE in three different samples: Pt (2.5 nm)/YIG (78 μm), Pt (2.5 nm)/YIG (21 nm), and Pt (2.5 nm)/BaM (1.2 μm). For each sample, three different experimental setup configurations were tested: illuminating from the sample's top, illuminating from the sample's bottom, and illuminating from both the top and bottom of the sample. The phenomena of PSVE in all cases were similar, with a difference that is no bigger than an order of magnitude. This confirms that the voltage is induced by PSVE instead of SEE. Only the sign of the voltage, but not its magnitude, flipped with the flipping of the magnetization of the MI film; this confirms the spin origin of the measured voltage. (All the PSVE information and experimental setup and discussion were taken from [48] with appropriate permissions).

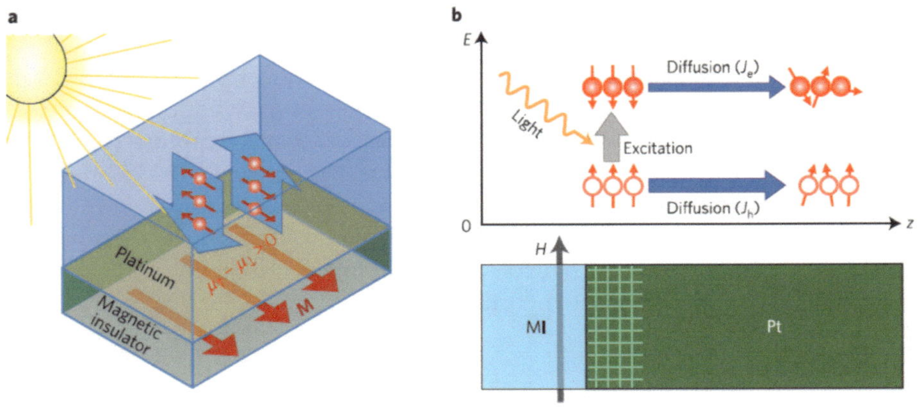

Figure 8.
(a) Photo-spin-voltaic effect in Pt/MI bilayer heterostructure. (b) Sketch of the physical mechanism underlying PSVE. When light illuminates the sample, photons excite electrons and generate nonequilibrium hot electrons and holes in the Pt atomic layers that are in proximity to the MI (the gridded region). The excited electrons and holes diffuse from Pt/MI interface to the Pt interface (along the +z direction), giving rise to spin currents (J_e and J_h). Source: [48], pp. 861, 865.

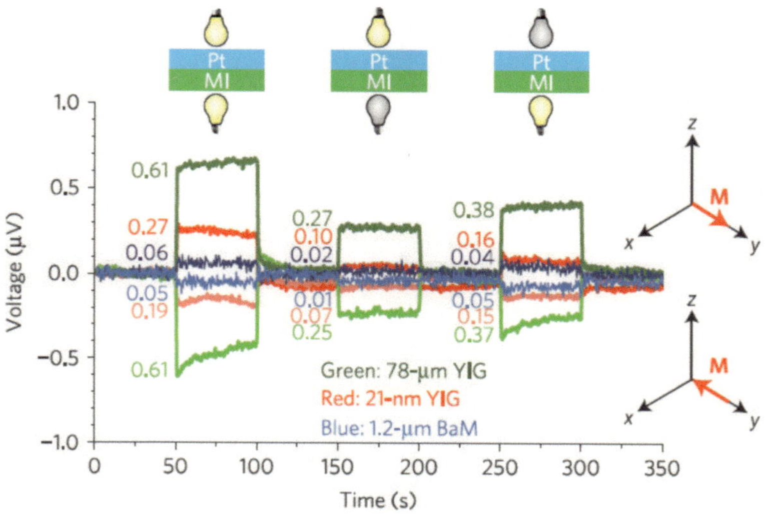

Figure 9.
Measurements for different illumination/magnetization configurations for three different samples Pt (2.5 nm)/YIG (78 μm), Pt (2.5 nm)/YIG (21 nm), and Pt (2.5 nm)/BaM (1.2 μm). Source: [48], p. 863.

3.3 Spin-orbit torque-assisted switching in magnetic insulators

The uniaxial anisotropy and the nonvolatile nature of easy axis-aligned magnetization within the BaM film can be used to design memory and logic-based systems. If the magnetization is up, it will keep its direction until a magnetic field flips it toward the opposite direction. If an efficient way can be found to switch the magnetization states of the magnetic insulator thin films, then they can be used in magnetic memory systems commercially [51].

In a NM/MI structure, such as Pt/BaM, SHE can be used to convert a charge current across the Pt surface into a spin current that flows across the thickness of Pt through spin-orbit coupling; this process will accumulate spins at the Pt/BaM interface. The spin accumulation generates spin-orbit torques (SOTs) that can be used to switch the BaM magnetization. Each electron provided by the charge current can undergo several spin-flip scatterings at the interface, breaking the conventional spin-torque switching limit and increasing the switching efficiency considerably [51].

We discuss the SOT experimental details of a Pt(5 nm)/BaM(3 nm) sample. The easy axis of the BaM film was perpendicular to the surface of the film. **Figure 10b** shows the hysteresis loop of the film, measured by a vibrating sample magnetometer, when an out-of-plane external magnetic field was applied (red curve). The olive curve shows the hysteresis loop along the hard axis when the external magnetic field is applied in the plane of the film. This figure confirms the perpendicular uniaxial anisotropy of the film, with a perpendicular anisotropy field of 17.6 kOe. A Hall bar structure was fabricated out of the Pt/BaM bilayer and is shown in **Figure 10a**. **Figure 10c** shows a hysteresis loop on the Hall resistance, revealing an anomalous Hall effect (AHE)-like behavior. It is unclear whether the AHE-like behavior is from magnetic proximity effect or spin Hall magnetoresistance. However, R_{AHE} behaves in a very similar manner to the perpendicular magnetization component of the BaM film $M_{\perp}$ (compare **Figure 10b** to **Figure 10c**). This allows for the gauging of $M_{\perp}$ in the BaM film by simply measuring R_{AHE}.

The first experiment demonstrated was the out-of-plane switching; the external magnetic field is fixed out of the film's plane and 20° off the easy axis. The purpose of this tilt was to break the magnetization symmetry due to the external field,

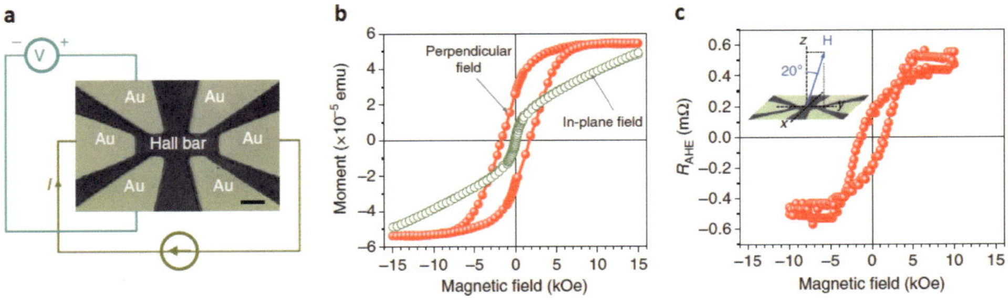

Figure 10.
(a) Optical image of the Pt (5 nm)/BaM (3 nm) Hall bar structure. (b) Magnetic hysteresis loops of the BaM film. (c) Anomalous Hall resistance R_{AHE} *of the Hall bar measured as a function of a magnetic field. The inset is a schematic showing the magnetic field H direction which is in the yz plane and 20 degrees away from the +z axis. Source: [51], p. 3.*

allowing for the observation of the SOT effect. One would expect that if the SOT field is along the *-z* direction, it would act against the external field, thereby increasing the total field required to saturate the magnetization within the BaM film, while a SOT field along the *z* direction will aid the external field, resulting in a smaller field required to saturate the magnetization of the BaM film.

Indeed, experimental results, shown in **Figure 11**, confirm exactly that. Namely, when charge currents of varying intensities are applied to the Pt film along the $-y$ direction, the SOT direction is opposite to that of H (as shown in **Figure 11a**), and the resultant hysteresis loops, gauged by R_{AHE}, become wider as the current intensity increases. This is shown in **Figure 11c**, where the gray loop is for $I = 0$; blue, $I = -2$ mA; olive, $I = -4$ mA; and red, $I = -6$ mA. This confirms the existence and

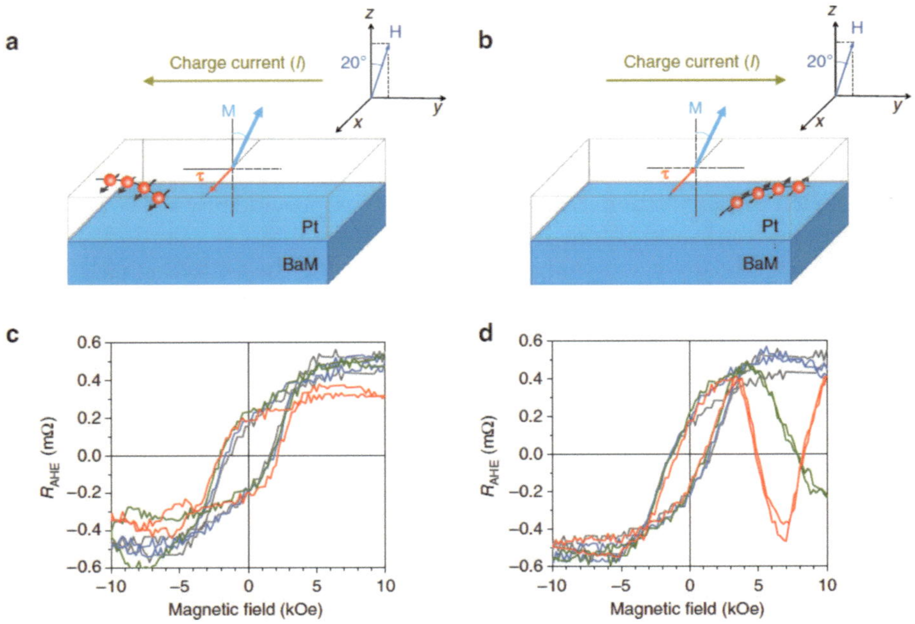

Figure 11.
Switching responses in Pt/BaM for out-of-plane magnetic fields. (a) and (b) Effects of charge currents I *in the Pt film on switching of the magnetization* M *in the BaM film under an out-of-plane field H. The red spheres with arrows represent spin-polarized electrons deflecting toward the BaM layer.* M *represents the magnetization of BaM. τ represents the spin torque due to SHE. The direction of* H *is indicated in the insert. (c) and (d) Anomalous Hall resistance* R_{AHE} *of the Hall bar measured as a function of a magnetic field for different charge currents. The field was applied 20 degrees away from the z axis, as shown in the insets of (a) and (b). In (c) gray,* I = 0; *blue,* I = −2 mA; *olive,* I = −4 mA; *and red,* I = −6 mA. *In (d) gray,* I = 0; *blue,* I = 2 mA; *olive,* I = 4mA; *and red,* I = 6 mA. *Source: [51], p. 4.*

the direction of pure spin current-generated SOTs near the interface, the magnitude of which is proportional to the intensity of the supplied current. Flipping the direction of the supplied charge currents flips the direction of the SOT as shown in **Figure 11b**. The resultant hysteresis loops, shown in **Figure 11d**, become narrower as the supplied charge current increases, indicating that SOT, in the direction of H, assisted in the magnetization flipping, reducing the overall total external field needed to flip $M_{\perp}$ in the BaM film.

Further experiments were performed to confirm the existence of spin current-generated SOT near the Pt/BaM interface. This time, the external field H was within the film plane. This means that applying a saturation field in the film plane will align the electron spins along the hard axis of the BaM film. When H is removed, the spins will return to their easy axis, randomly up or down, resulting in a net $M_{\perp}$ of zero. However, when a charge current is supplied to the Pt surface, SOT near the interface will influence the direction of the spins of electrons when H is reduced and they start to align to their easy axis. This results in remnant magnetization that is represented by a hysteresis loop. If the direction of the supplied charge current is flipped, the resultant hysteresis loop will be the opposite of the first hysteresis loop. This is shown in **Figure 12**. This confirms that the direction of the SOT can be controlled by changing the sign of the supplied charge current.

These results confirm that SOT due to pure spin currents, generated by SHE in Pt/BaM structures, can be used to assist the magnetization switching in BaM films. It should be noted however, that SHE generates two different torques: a damping-like torque (DLT) and a field-like torque (FLT). The effective fields for DLT and FLT are H_{DLT} and H_{FLT}, respectively. Thus, the total field affecting $M_{\perp}$ of the BaM film can be written as follows [51]:

$$H_{\mathrm{total}} = H + H_{\mathrm{a}} + H_{\mathrm{FLT}}x + H_{\mathrm{DLT}}(m \times x) \quad (5)$$

where H is the external field as indicated in **Figure 11**, H_{a} is the anisotropy field of the BaM film, and x is the unit vector along the $+x$ direction. Two different simulation models were carried out to determine the SOT field strength: macrospin model simulation and microspin model simulation.

Carrying out both simulations involved three main steps: first, H_{c} was calculated when J_{c} is set to zero. H_{DLT} and H_{FLT} were both set to zero as well. H_{a} was set such that when H is equal to the experimentally measured H_{c} and pointing in the direction opposite to its initial direction, m flips. The second step considers the case when

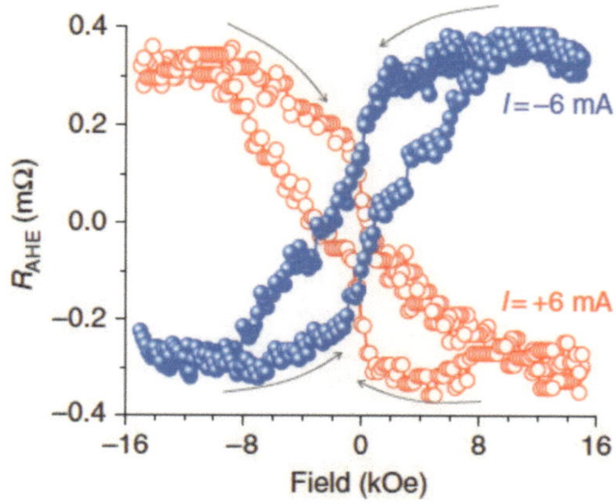

Figure 12.
Anomalous Hall resistance R_{AHE} measured as a function of a magnetic field along the y axis for I= +6 mA and I= −6 mA, respectively. Source: [51], p. 5.

$J_c \neq 0$ and $H_{FLT} = 0$. Simulations are then performed at given H_{DLT} values to find the corresponding H_c values. This is because a flip in the direction of H_{DLT} breaks the symmetry, thereby affecting H_c. A flip in the direction of H_{FLT} on the other hand does not affect H_c as H_{FLT} is orthogonal to H_a, H, and m. The third step considers non-zero values of H_{DLT} and repeats the simulations to find H_c for given combinations of H_{DLT} and H_{FLT} values.

The results from running the two different models of simulations were very close and are shown in **Figure 13**. The blue dots show the linear nature of the relationship between H_c and H_{DLT}, when $H_{FLT} = 0$. This is similar to the experimental H_c vs. J_c data. The simulation showed that when H_{DLT} is -400 Oe, H_c increases to about 2.0 kOe, and when H_{DLT} is 400 Oe, H_c decreases to about 0.95 kOe. This same change was experimentally observed when J_c changed between -10^7 A cm^{-2} and 10^7 A cm^{-2}. Thus, we can conclude that H_{DLT} in the Pt/BaM is about 400 Oe at $J_c = 10^7$ A cm^{-2}. The red and olive dots in **Figure 13a** and **b** show the same relationship when $H_{FLT} = H_{DLT}/2$ and $H_{FLT} = H_{DLT}$, respectively. The red dots show that the effect of H_{FLT} is negligible when $H_{FLT} = H_{DLT}/2$, while the olive dots show a deviation for strong negative charge currents that was not observed experimentally. The red and olive portions of both figures prove that the majority of the SHE generated torque is due to DLT, with FLT having a relatively small effect in comparison. The experimental results, along with the simulation data, show that SOT in Pt/BaM films can reduce the required switching field by as much as 500 Oe.

Further improvements and enhancements in the switching efficiency can be achieved by using materials with higher spin-orbit coupling, resulting in stronger SOT. Topological insulators exhibit such requirements and will be the topic of the next section. (All figures, experimentation setup, and results were taken from [51] with appropriate permissions).

3.4 Magnetization switching with topological insulators

Topological insulators (TI) are of great interest in spintronic-related studies. A TI is a material with nontrivial symmetry-protected topological order that behaves as an insulator in its interior but whose surface contains conducting states. What differentiates a TI from other materials with conducting surfaces is that its surface

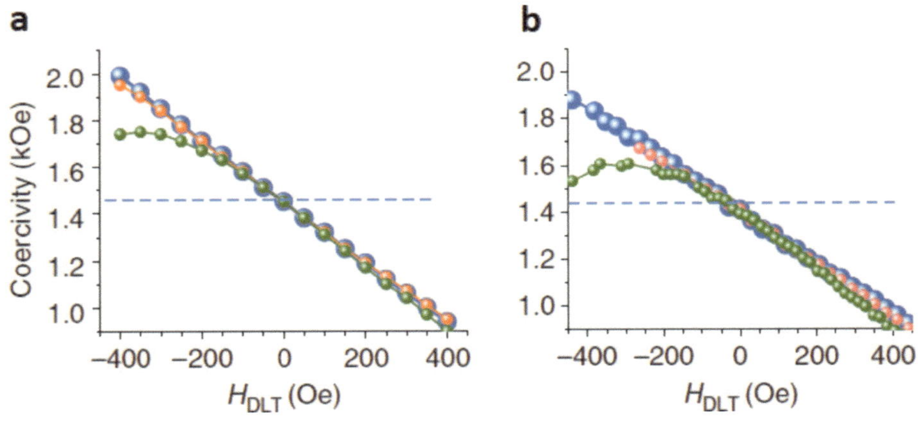

Figure 13.
(a) and (b) Coercivity vs. DLT field (H_{DLT}) estimated for three different field-like torque (FLT) fields (H_{FLT}) through macrospin and full micromagnetic simulations, respectively. Large blue spheres, $H_{FLT} = 0$; small red spheres, $H_{FLT} = H_{DLT}/2$; and small olive spheres, $H_{FLT} = H_{DLT}$. The dash line in (a) and (b) is the H_c at $I = 0$. All the measurements were done at room temperature. Source: [51], p. 4.

states are time-reversal symmetry-protected. Due to the very strong spin-orbit coupling of TIs [10, 52], if a charge current is supplied to their surface, the surface states induce spin polarity and therefore generate a spin current, owing to the SHE. The SHE in TIs is several times stronger than in heavy metals such as Pt, and it can become hundreds of times stronger at very lower temperatures [10].

Theoretically, the very strong SHE in a TI can generate SOT that is much stronger than its counterpart in heavy metals. This strong SOT can then be exploited for magnetization switching by pairing it with a ferromagnet, similar to what was discussed in the previous section. Using a conductive ferromagnet, however, can completely suppress the surface states of a TI [49–56], preventing the generation of spin currents, therefore making it impossible for SOT magnetization switching to happen in TI/conductive ferromagnet structures.

Here, the usefulness and importance of magnetic insulators are again emphasized. Pairing a TI with MI keeps the integrity of the surface states. Various materials can be used to create a TI, such as $(Bi_xSb_{1\text{-}x})_2\,Te_3$. The choice of x can ensure protection from time-reversal symmetry. **Figure 14a** and **b** show the sheet resistance measurements of a $(Bi_xSb_{1\text{-}x})_2Te_3$ film that was grown on a MI ($Tm_3Fe_5O_{12}$, TIG) for x = 0.2 and x = 0.3, respectively. The lower inset of both figures show the broken symmetry of the topological surface states of both configurations. The sheet resistance in both figures shows a linear portion, attributed to the normal Hall effect, and a hysteresis loop portion. The different slopes indicate opposite carriers in each sample. The hysteresis portion indicates strong magnetic uniaxial anisotropy in the TI owing to highly spin-polarized electrons on the TI's surface. This uniaxial anisotropy is maintained at room temperature and up to T = 400 K [57].

In another experiment, the authors used a Bi_2Se_3/BaM heterostructure to explore the effect of topological surface state in switching the magnetization of a magnetic insulator [10]. The BaM layer used had similar characteristics to the BaM layer used in the Pt/BaM experiment. The BaM film was 5-nm-thick and had a uniaxial anisotropy axis perpendicular to the surface, as shown by the two hysteresis loops in **Figure 15a**. The blue hysteresis loop was measured when the external field was applied perpendicular to the BaM film's surface. The red loop was measured when an external field was applied along the BaM film plane. The two loops together confirm the perpendicular orientation of the anisotropy axis of the BaM film.

A Hall bar was fabricated on the Bi_2Se_3/BaM bilayer film. **Figure 15c** shows that, similar to the Hall bar setup of the Pt/BaM experiment discussed in the previous section, the AHE contribution to the Hall bar resistance, R_{AHE}, scales with the

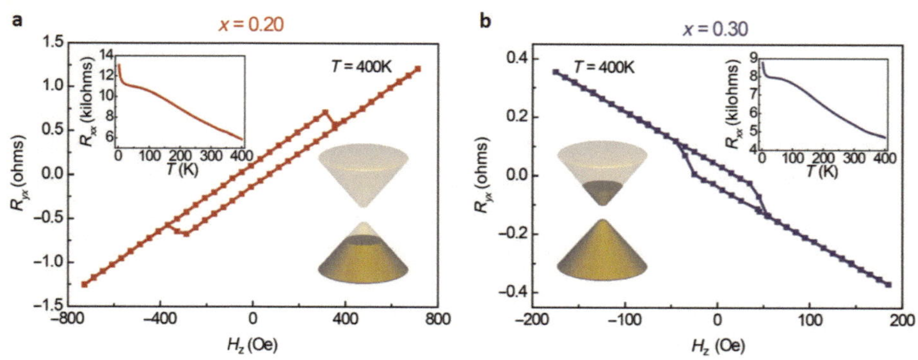

Figure 14.
(a) and (b) Hall traces of TIG/$(Bi_xSb_{1-x})_2\,Te_3$ for x = 0.20 and 0.30, respectively. The upper insets show the corresponding temperature dependence of R_{xx}. The lower insets show schematic drawings of the corresponding chemical potential position. Source: [57], p. 2.

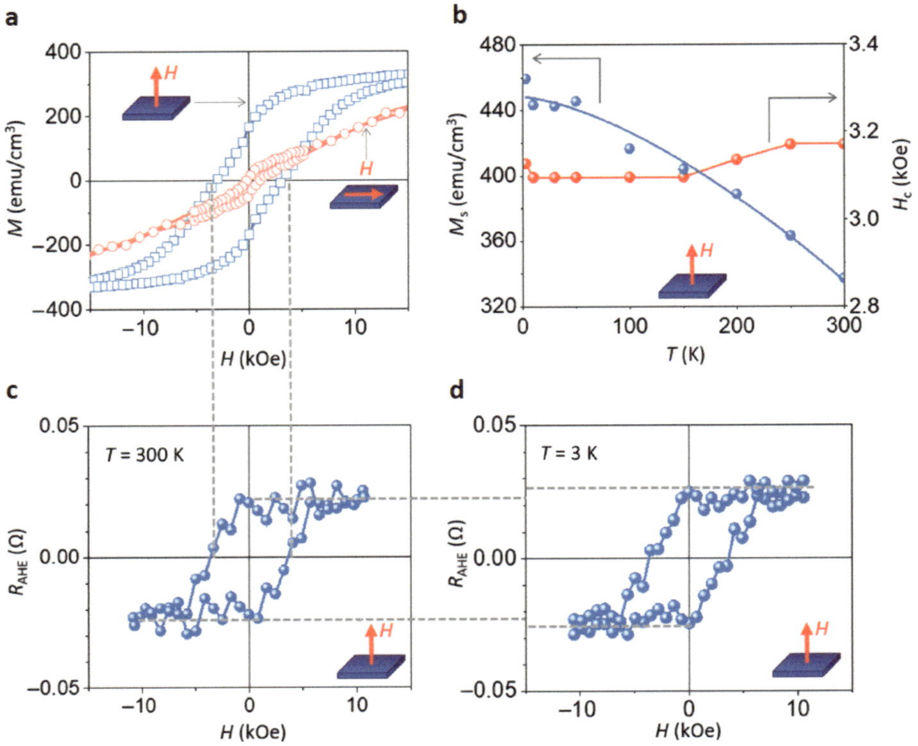

Figure 15.
(a) Magnetization (M) vs. field (H) loops for the $Bi_2Se_3/BaFe_{12}O_{19}$ *sample. (b) Saturation magnetization* (M_s) *and coercive field* (H_c) *as a function of T. (c) and (d)* R_{AHE} *vs. field (H) loops measured at T = 300 K and T = 3 K. Source: [10], p. 4.*

perpendicular magnetization $M_{\perp}$ of the BaM film and therefore can be used as an easy way to probe $M_{\perp}$ of the BaM film during experimentation. While **Figure 15c** shows the R_{AHE} response in room temperature settings, **Figure 15d** shows R_{AHE} when T = 3 K. The figure shows a hysteresis loop that is very close to the one in **Figure 15c**. The hysteresis loop widths are very similar, indicating that the same field strength *H* is required to saturate the magnetization of the BaM film when T = 3 K and when T = 300 K. The value of R_{AHE} is slightly higher when T = 3 K than when T = 300 K, indicating that the saturation of the BaM film increases slightly as T decreases. The effect of the temperature change on the values of the saturation magnetization of the BaM film and the coercive field required to saturate it are both shown in **Figure 15b**.

Figure 16a shows the SOT switching experiment configuration. An external field *H* was applied along the *x* direction to aid in the SOT switching of $M_{\perp}$ in the BaM film. $M_{\perp}$ was initially along the positive *z* direction. **Figure 16b** shows the experimentation results when T = 200 K and *H* = -15 kOe. The blue data points show the effect of sweeping a DC charge current I_{dc} from positive to negative. When the supplied current was positive, the SOT direction was also in the positive *z* direction, so we see no change in R_{AHE}. When I_{dc} is <0, however, the SOT direction flips to the negative *z* direction. When the torque generated by the polarized spin accumulation at the Bi_2Se_3/BaM becomes strong enough, the magnetization within the BaM film switches to the negative *z* direction, as indicated by the change in R_{AHE} value from positive to negative. Sweeping I_{dc} from negative to positive generates SOT in the +z direction when $I_{dc} > 0$; when the SOT is strong enough, it flips the magnetization of the BaM film again, indicated by the red points in the figure. The result from both cases is a hysteresis loop in R_{AHE} that can be seen in the figure.

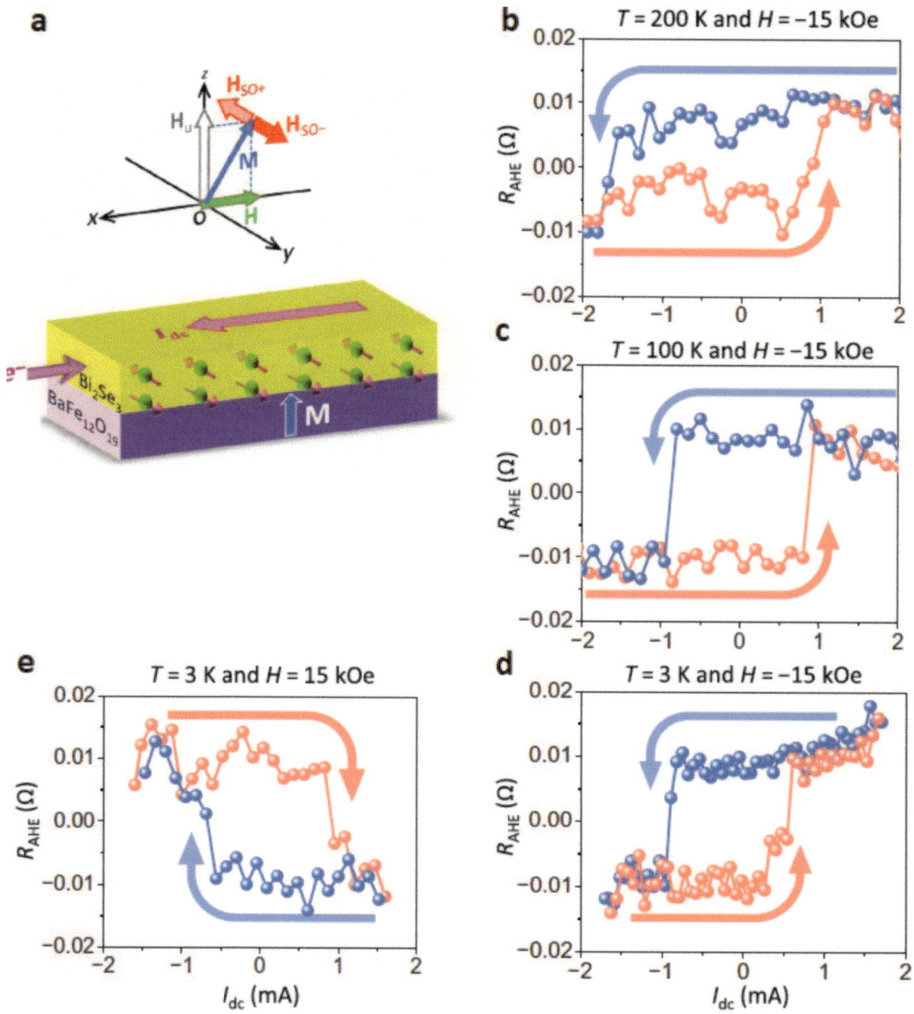

Figure 16.
SOT-induced switching in Bi_2Se_3/BaM. (a) Experimental configuration. (b to e) AHE resistance (R_{AHE}) measured as a function of charge current (I_{dc}) at different fields (H) and temperatures (T), as indicated. The arrows in (b) to (e) indicate the current sweeping directions. Source: [10], p. 5.

Figure 16c, d, and **e** shows the results of the same experiment performed at decreasing temperatures. The figures clearly indicate that the current required for magnetization switching becomes smaller as temperature decreases. This is due to the enhancement of the topological surface states in Bi_2Se_3 as T decreases.

Figure 17 further demonstrates the effect of SOT on the magnetization switching of the BaM film. The experiment was performed at T = 3 K; the external field was applied at 45 degrees angle out of the plane of the film as shown in the inset of the figure. The blue hysteresis loop is the result of applying a negative charge current that generated a SOT acting against *H*. The result is a wider hysteresis loop when compared with the normal hysteresis loop of the BaM film shown in **Figure 15**. This is due to the SOT acting against *H*, therefore hindering the magnetization switching and requiring a stronger external field to switch the magnetization of the BaM film. The red hysteresis loop shows the result of applying a positive charge current, which caused SOT that was in the direction of the external magnetic field, significantly decreasing the switching field required as shown by the much narrower hysteresis loop. This confirms the strength and significance of SOT in TIs and how it can be used to assist in magnetization switching.

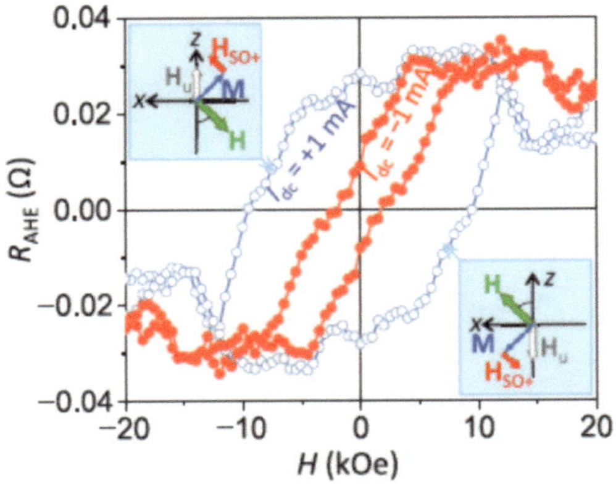

Figure 17.
Effects of I_{dc} on R_{AHE} hysteresis loops at T = 3 K in Bi_2Se_3/BaM. Source: [10], p. 6.

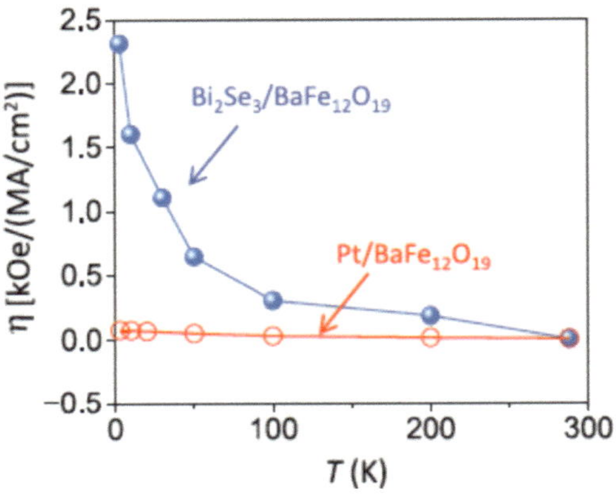

Figure 18.
SOT efficiency (η) as a function of T in Bi_2Se_3/BaM and Pt/BaM. The data were all measured at a field applied at an angle of 45 degrees away from the film normal direction. The data on Pt/BaM were measured with a Hall bar structure that had the same dimension as the Bi_2Se_3/BaM Hall bar. Source: [10], p. 6.

The efficiency of SOT switching can be calculated using the following expression [58]:

$$\eta = \frac{H_{SW}(I_{dc} > 0) - H_{SW}(I_{dc} < 0)}{2|I_{dc}|/(wt)} \tag{6}$$

where H_{SW} is the switching field, w is the Hall bar width, and t is Bi_2Se_3 or Pt thickness. The increase of SOT efficiency as the temperature decreases is demonstrated in **Figure 18**. The blue data points show η of Bi_2Se_3/BaM as a function of temperature. Note that the Debye temperature of Bi_2Se_3 is about 180 K [59]. When T is > > Debye, the efficiency is proportional to T^{-1}, but when the temperature is < < Debye (at T = 100 K), the efficiency is proportional to T^{-5}. Pt/BaM SOT efficiency is also shown as a function of temperature on the same plot (red points). The data show that decreasing the temperature has a negligible effect on the efficiency in Pt/BaM bilayers. The exponential improvement in SOT efficiency of Bi_2Se_3/BaM is caused by the fact that the surface conductance in TIs increases with decreasing temperatures, while the bulk conductance decreases. This means that the decrease in T enhances the TSS in TIs, resulting in a higher charge current to spin current conversion efficiency, increasing the spin polarity in TIs and therefore generating higher SOT. (All figures, experimentation setup and results were taken from [10] with appropriate permissions).

3.5 Summary and outlook

Magnetic insulators with perpendicular anisotropy have become an important class of materials in the development of spintronic devices. For magnetic domain devices, the low-damping and large anisotropy features can enable high-speed domain-wall motion with a small current threshold, fueling the development of domain-wall memory and logic devices. Moreover, low-damping is significant for SOT oscillator applications, where the current threshold for self-oscillations decreases with damping. Recent experiments show that spin waves can be used to control magnetic domains through spin-orbit torques [60, 61]; this effect can be amplified and become more efficient in magnetic insulators. The strong magnetic anisotropy also allows the engineering of spin-wave dispersion relation without the need for large bias magnetic fields [62]. This will expand the horizon for magnonic and spin-wave devices, allowing the development of new magnon-photon coupling devices for quantum transduction and microwave photonic systems [63, 64].

Author details

Laith Alahmed and Peng Li*
Department of Electrical and Computer Engineering, Auburn University, Auburn, USA

*Address all correspondence to: peng.li@auburn.edu

References

[1] Wolf SA, Awschalom DD, Buhrman RA, Daughton JM, von Molnár S, Roukes ML, et al. Spintronics: A spin-based electronics vision for the future. Science. 2001;**294**(5546): 1488-1495. Available from: https://science.sciencemag.org/content/294/5546/1488

[2] I. Žutić, J. Fabian, and S. Das Sarma, Spintronics: Fundamentals and applications. Reviews of Modern Physics. 2004;**76**:323-410. DOI: 10.1103/RevModPhys.76.323

[3] Fert A. Nobel lecture: Origin, development, and future of spintronics. Reviews of Modern Physics. 2008;**80**: 1517-1530. DOI: 10.1103/RevModPhys.80.1517

[4] Ikeda S, Miura K, Yamamoto H, Mizunuma K, Gan HD, Endo M, et al. A perpendicular-anisotropy CoFeB-MgO magnetic tunnel junction. Nature Materials. 2010;**9**(9):721-724. DOI: 10.1038/nmat2804

[5] Harris VG. Modern microwave ferrites. IEEE Transactions on Magnetics. March 2012;**48**(3):1075-1104

[6] Pullar RC. Hexagonal ferrites: A review of the synthesis, properties and applications of hexaferrite ceramics. Progress in Materials Science. 2012; **57**(7):1191-1334. Available from: http://www.sciencedirect.com/science/article/pii/S0079642512000369

[7] Malkinski L. Advanced Magnetic Materials. Croatia: IntechOpen; 2012

[8] Carosella CA, Chrisey DB, Lubitz P, Horwitz JS, Dorsey P, Seed R, et al. Pulsed laser deposition of epitaxial $BaFe_{12}O_{19}$ thin films. Journal of Applied Physics. 1992;**71**(10):5107-5110. DOI: 10.1063/1.350614

[9] Song Y-Y, Kalarickal S, Patton CE. Optimized pulsed laser deposited barium ferrite thin films with narrow ferromagnetic resonance linewidths. Journal of Applied Physics. 2003;**94**(8): 5103-5110. DOI: 10.1063/1.1608475

[10] Li P, Kally J, Zhang SS-L, Pillsbury T, Ding J, Csaba G, et al. Magnetization switching using topological surface states. Science Advances. 2019;**5**(8). Available from: https://advances.sciencemag.org/content/5/8/eaaw3415

[11] Geiler AL, Yang A, Zuo X, Yoon SD, Chen Y, Harris VG, et al. Atomic scale design and control of cation distribution in hexagonal ferrites. Physical Review Letters. 2008;**101**:067201. DOI: 10.1103/PhysRevLett.101.067201

[12] Mohebbi M, Ebnabbasi K, Vittoria C. In-situ deposition of c-axis oriented barium ferrite films for microwave applications. IEEE Transactions on Magnetics. July 2013; **49**(7):4207-4209

[13] Liu H, Avrutin V, Xiao B, Rowe E, Liu H, Özgür, et al. Epitaxial relationship of mbe grown barium hexaferrite (0001) films on sapphire (0001). Journal of Crystal Growth. 2010;**312**(5):671-675. Available from: http://www.sciencedirect.com/science/article/pii/S0022024809010963

[14] Kranov YA, Abuzir A, Prakash T, McIlroy DN, Yeh WJ. Barium hexaferrite thick films made by liquid phase epitaxy reflow method. IEEE Transactions on Magnetics. October 2006;**42**(10):3338-3340

[15] Yoon SD, Vittoria C. Microwave and magnetic properties of barium hexaferrite films having the c-axis in the film plane by liquid phase epitaxy technique. Journal of Applied Physics. 2003;**93**(10):8597-8599. DOI: 10.1063/1.1557791

[16] Cho TS, Je JH, Noh DY. Formation of crystalline Ba-ferrite phase from -

Fe_2O_3 phase in amorphous precursor. Applied Physics Letters. 2000;**76**(3): 303-305. DOI: 10.1063/1.125727

[17] Zhang X, Meng S, Song D, Zhang Y, Yue Z, Harris VG. Epitaxially grown BaM hexaferrite films having uniaxial axis in the film plane for self-biased devices. Scientific Reports. 2017;7(1): 44193. DOI: 10.1038/srep44193

[18] Li P, Ellsworth D, Chang H, Janantha P, Richardson D, Shah F, et al. Generation of pure spin currents via spin Seebeck effect in self-biased hexagonal ferrite thin films. Applied Physics Letters. 2014;**105**(24):242412. DOI: 10.1063/1.4904479

[19] Uchida K, Ishida M, Kikkawa T, Kirihara A, Murakami T, Saitoh E. Longitudinal spin Seebeck effect: From fundamentals to applications. Journal of Physics: Condensed Matter. 2014; **26**(34):343202. DOI: 10.1088

[20] Uchida K, Xiao J, Adachi H, Ohe J, Takahashi S, Ieda J, et al. Spin Seebeck insulator. Nature Materials. 2010; **9**(11):894-897. DOI: 10.1038/nmat2856

[21] Adachi H, Uchida K-I, Saitoh E, Ohe J-I, Takahashi S, Maekawa S. Gigantic enhancement of spin Seebeck effect by phonon drag. Applied Physics Letters. 2010;**97**(25):252506. DOI: 10.1063/1.3529944

[22] Uchida K-I, Nonaka T, Ota T, Saitoh E. Longitudinal spin-Seebeck effect in sintered polycrystalline (mn, zn)Fe_2O_4. Applied Physics Letters. 2010;**97**(26):262504. DOI: 10.1063/1.3533397

[23] Qu D, Huang SY, Hu J, Wu R, Chien CL, et al. Physical Review Letters. 2013;**110**:067206. DOI: 10.1103/PhysRevLett.110.067206

[24] Urban R, Woltersdorf G, Heinrich B. Gilbert damping in single and multilayer ultrathin films: Role of interfaces in nonlocal spin dynamics. Physical Review Letters. 2001;**87**:217204. DOI: 10.1103/PhysRevLett.87.217204

[25] Mizukami S, Ando Y, Miyazaki T. Effect of spin diffusion on gilbert damping for a very thin permalloy layer in Cu/permalloy/Cu/Pt films. Physical Review B. 2002;**66**:104413. DOI: 10.1103/PhysRevB.66.104413

[26] Tserkovnyak Y, Brataas A, Bauer GEW. Enhanced gilbert damping in thin ferromagnetic films. Physical Review Letters. 2002;**88**:117601. DOI: 10.1103/PhysRevLett.88.117601

[27] Heinrich B, Burrowes C, Montoya E, Kardasz B, Girt E, Song Y-Y, et al. Spin pumping at the magnetic insulator (YIG)/normal metal (Au) interfaces. Physical Review Letters. 2011;**107**: 066604. DOI: 10.1103/PhysRevLett.107.066604

[28] Uchida K, Takahashi S, Harii K, Ieda J, Koshibae W, Ando K, et al. Observation of the spin Seebeck effect. Nature. 2008;**455**(7214):778-781. DOI: 10.1038/nature07321

[29] Uchida K, Ota T, Harii K, Takahashi S, Maekawa S, Fujikawa Y, et al. Spin-Seebeck effects in $Ni_{81}Fe_{19}$/Pt films. Solid State Communications. 150, 2010;**11**:524-528. Spin Caloritronics. Available from: http://www.sciencedirect.com/science/article/pii/S003810980900698X

[30] Jaworski CM, Yang J, Mack S, Awschalom DD, Heremans JP, Myers RC. Observation of the spin-Seebeck effect in a ferromagnetic semiconductor. Nature Materials. 2010; **9**(11):898-903. DOI: 10.1038/nmat2860

[31] Jaworski CM, Myers RC, Johnston-Halperin E, Heremans JP. Giant spin Seebeck effect in a non-magnetic material. Nature. 2012;**487**(7406): 210-213. DOI: 10.1038/nature11221

[32] Meier D, Kuschel T, Shen L, Gupta A, Kikkawa T, Uchida K, et al. Thermally driven spin and charge currents in thin $NiFe_2O_4$/Pt films. Physical Review B. February 2013;**87**:054421. DOI: 10.1103/PhysRevB.87.054421

[33] Ramos R, Kikkawa T, Uchida K, Adachi H, Lucas I, Aguirre MH, et al. Observation of the spin Seebeck effect in epitaxial fe3o4 thin films. Applied Physics Letters. 2013;**102**(7):072413. DOI: 10.1063/1.4793486

[34] Ashcroft N, Mermin D. Solid State Physics. Belmont, California: Brooks/Cole; 1976

[35] Bosu S, Sakuraba Y, Uchida K, Saito K, Ota T, Saitoh E, et al. Spin Seebeck effect in thin films of the heusler compound Co_2MnSi. Physical Review B. June 2011;**83**:224401. DOI: 10.1103/PhysRevB.83.224401

[36] Slonczewski JC. Conductance and exchange coupling of two ferromagnets separated by a tunneling barrier. Physical Review B. April 1989;**39**: 6995-7002. DOI: 10.1103/PhysRevB.39.6995

[37] Jedema FJ, Filip AT, van Wees BJ. Electrical spin injection and accumulation at room temperature in an all-metal mesoscopic spin valve. Nature. 2001; **410**(6826):345-348. DOI: 10.1038/35066533

[38] Maekawa S, Adachi H, Uchida K-I, Ieda J, Saitoh E. Spin current: Experimental and theoretical aspects. Journal of the Physical Society of Japan. 2013;**82**(10):102002. DOI: 10.7566/JPSJ.82.102002

[39] Uchida K, Ota T, Adachi H, Xiao J, Nonaka T, Kajiwara Y, et al. Thermal spin pumping and magnon-phonon-mediated spin-Seebeck effect. Journal of Applied Physics. 2012;**111**(10):103903. DOI: 10.1063/1.4716012

[40] Uchida K-I, Adachi H, Ota T, Nakayama H, Maekawa S, Saitoh E. Observation of longitudinal spin-Seebeck effect in magnetic insulators. Applied Physics Letters. 2010;**97**(17): 172505. DOI: 10.1063/1.3507386

[41] Berger L. Application of the side-jump model to the Hall effect and Nernst effect in ferromagnets. Physical Review B. 1972;**5**:1862-1870. DOI: 10.1103/PhysRevB.5.1862

[42] Miyasato T, Abe N, Fujii T, Asamitsu A, Onoda S, Onose Y, et al. Crossover behavior of the anomalous Hall effect and anomalous Nernst effect in itinerant ferromagnets. Physical Review Letters. August 2007;**99**:086602. DOI: 10.1103/PhysRevLett.99.086602

[43] Saitoh E, Ueda M, Miyajima H, Tatara G. Conversion of spin current into charge current at room temperature: Inverse spin-Hall effect. Applied Physics Letters. 2006;**88**(18): 182509. DOI: 10.1063/1.2199473

[44] Kimura T, Otani Y, Sato T, Takahashi S, Maekawa S. Room-temperature reversible spin Hall effect. Physical Review Letters. 2007;**98**:156601. DOI: 10.1103/PhysRevLett.98.156601

[45] Azevedo A, Vilela Leão LH, Rodriguez-Suarez RL, Oliveira AB, Rezende SM. dc effect in ferromagnetic resonance: Evidence of the spin-pumping effect? Journal of Applied Physics. 2005; **97**(10):10C715. DOI: 10.1063/1.1855251

[46] Valenzuela SO, Tinkham M. Direct electronic measurement of the spin Hall effect. Nature. 2006;**442**(7099):176-179. DOI: 10.1038/nature04937

[47] Costache MV, Sladkov M, Watts SM, van der Wal CH, van Wees BJ. Electrical detection of spin pumping due to the processing magnetization of a single ferromagnet. Physical Review Letters. 2006;**97**:216603. DOI: 10.1103/PhysRevLett.97.216603

[48] Ellsworth D, Lu L, Lan J, Chang H, Li P, Wang Z, et al. Photo-spin-voltaic

effect. Nature Physics. 2016;**12**(9): 861-866. DOI: 10.1038/nphys3738

[49] Chen Y-J, Huang S-Y. Light-induced thermal spin current. Physical Review B. 2019;**99**:094426. DOI: 10.1103/PhysRevB.99.094426

[50] Kattel S, Murphy JR, Ellsworth D, Ding J, Liu T, Li P, et al. Broadband optical detection using the spin Seebeck effect. Physical Review Applied. 2019; **12**:034047. DOI: 10.1103/ PhysRevApplied.12.034047

[51] Li P, Liu T, Chang H, Kalitsov A, Zhang W, Csaba G, et al. Spin-orbit torque-assisted switching in magnetic insulator thin films with perpendicular magnetic anisotropy. Nature Communications. 2016;7(1):12688. DOI: 10.1038/ncomms12688

[52] Zhang J, Velev JP, Dang X, Tsymbal EY. Band structure and spin texture of Bi_2Se_3*3d* ferromagnetic metal interface. Physical Review B. 2016;**94**: 014435. DOI: 10.1103/ PhysRevB.94.014435

[53] Avci CO, Quindeau A, Pai C-F, Mann M, Caretta L, Tang AS, et al. Current-induced switching in a magnetic insulator. Nature Materials. 2017;**16**(3): 309-314. DOI: 10.1038/nmat4812

[54] Hoffmann A. Spin Hall effects in metals. IEEE Transactions on Magnetics. 2013;**49**(10):5172-5193

[55] Zhang S. Spin Hall effect in the presence of spin diffusion. Physical Review Letters. 2000;**85**:393-396. DOI: 10.1103/PhysRevLett.85.393

[56] Day C. Two groups observe the spin Hall effect in semiconductors. Physics Today. 2005;**58**(2):17-19. DOI: 10.1063/ 1.1897513

[57] Tang C, Chang C-Z, Zhao G, Liu Y, Jiang Z, Liu C-X, et al. Above 400-k robust perpendicular ferromagnetic phase in a topological insulator. Science Advances. 2017;**3**:6. Available from: https://advances.sciencemag.org/content/3/6/e1700307

[58] Khang NHD, Ueda Y, Hai PN. A conductive topological insulator with large spin Hall effect for ultralow power spin-orbit torque switching. Nature Materials. 2018;**17**(9):808-813. DOI: 10.1038/s41563-018-0137-y

[59] Chen X, Zhou HD, Kiswandhi A, Miotkowski I, Chen YP, Sharma PA, et al. Thermal expansion coefficients of Bi_2Se_3 and Sb_2Te_3 crystals from 10 k to 270 k. Applied Physics Letters. 2011; **99**(26):261912. DOI: 10.1063/1.3672198

[60] Wang Y, Zhu D, Yang Y, Lee K, Mishra R, Go G, et al. Magnetization switching by magnon-mediated spin torque through an antiferromagnetic insulator. Science. 2019;**366**(6469):1125. Available from: http://science.sciencemag.org/content/366/6469/1125.abstract

[61] Han J, Zhang P, Hou JT, Siddiqui SA, Liu L. Mutual control of coherent spin waves and magnetic domain walls in a magnonic device. Science. 2019;**366**(6469):1121-1125. Available from: https://science.sciencemag.org/content/366/6469/1121

[62] Soumah L, Beaulieu N, Qassym L, Carrétéro C, Jacquet E, Lebourgeois R, et al. Ultra-low damping insulating magnetic thin films get perpendicular. Nature Communications. 2018;**9**(1): 3355. DOI: 10.1038/s41467-018-05732-1

[63] Hou JT, Liu L. Strong coupling between microwave photons and nanomagnet magnons. Physical Review Letters. 2019;**123**:107702. DOI: 10.1103/ PhysRevLett.123.107702

[64] Li Y, Polakovic T, Wang Y-L, Xu J, Lendinez S, Zhang Z, et al. Strong coupling between magnons and microwave photons in on-chip ferromagnet-superconductor thin-film devices. Physical Review Letters. 2019; **123**:107701. DOI: 10.1103/ PhysRevLett.123.107701

Section 2

Heusler Compounds

Chapter 4

Magnetic Full-Heusler Compounds for Thermoelectric Applications

Kei Hayashi, Hezhang Li, Mao Eguchi,
Yoshimi Nagashima and Yuzuru Miyazaki

Abstract

Full-Heusler compounds exhibit a variety of magnetic properties such as non-magnetism, ferromagnetism, ferrimagnetism and anti-ferromagnetism. In recent years, they have attracted significant attention as potential thermoelectric (TE) materials that convert thermal energy directly into electricity. This chapter reviews the theoretical and experimental studies on the TE properties of magnetic full-Heusler compounds. In Section 1, a brief outline of TE power generation is described. Section 2 introduces the crystal structures and magnetic properties of full-Heusler compounds. The TE properties of full-Heusler compounds are presented in Sections 3 and 4. The relationship between magnetism, TE properties and order degree of full-Heusler compounds is elaborated.

Keywords: full-Heusler compounds, half-metal, spin-gapless semiconductor, thermoelectric properties, order degree

1. Introduction

Thermoelectric (TE) power generation using TE devices is one of the key technologies to solve global energy problem, owing to its availability of direct conversion of thermal energy into electricity [1–3]. A schematic figure of a TE device is shown in **Figure 1**. It consists of n- and p-type TE materials connected in series electrically with metal electrodes and arranged thermally in parallel. The TE materials are wedged between ceramic plates. When one side of the device is heated and the other side is cooled, electrons and holes in the n- and p-type TE materials, respectively, diffuse from the hot side to the cold side, thus generating a flow of electric current.

To commercialise TE devices, there is a need to improve their TE efficiency. The maximum TE efficiency, η_{max}, is an increasing function of the dimensionless figure-of-merit, zT, expressed as:

$$\eta_{\text{max}} = \frac{T_{\text{H}} - T_{\text{C}}}{T_{\text{H}}} \frac{\sqrt{1+zT} - 1}{\sqrt{1+zT} + T_{\text{C}}/T_{\text{H}}}, \tag{1}$$

where T_{H} and T_{C} are the heating and cooling temperature, respectively. The dimensionless figure-of-merit, zT, is determined by TE properties (S: Seebeck

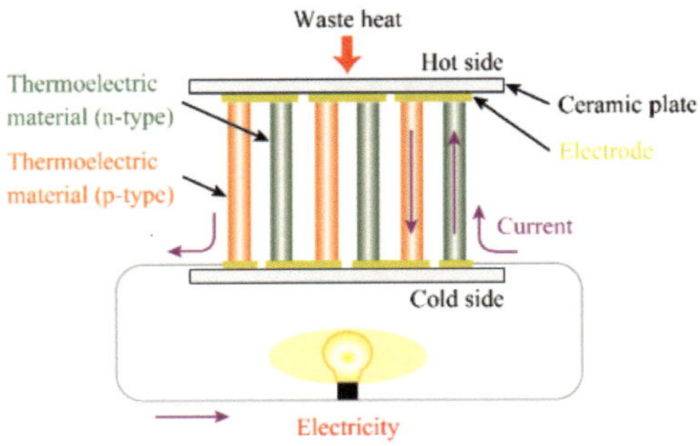

Figure 1.
Schematic figure of a thermoelectric (TE) power generation device.

coefficient, σ: electrical conductivity, κ: thermal conductivity) of the individual TE materials in the device.

$$zT = \frac{S^2\sigma}{\kappa}T, \tag{2}$$

where T is the absolute temperature. The product $S^2\sigma$ is called the power factor (PF), which is a measure of electric power generated using the TE material. To achieve high TE efficiency (standard levels for practical use are $zT > 1$ and $\mathrm{PF} > 2 \times 10^{-3}$ W/K^{2}m), high S, high σ and low κ are required. To meet these requirements, a variety of TE materials have been explored, such as chalcogenides, skutterudites, clathrates, silicides, Zintl compounds, half-Heusler compounds and oxides [1–3]. Most of these materials are semiconductors because in general they have high S than metals. However, recent theoretical and experimental studies have revealed that metals, in particular, half-metallic full-Heusler compounds have relatively high S as well as high σ. In addition, their junction with a metal electrode is robust compared to that of semiconductors, which is also an advantage.

In Section 2, the crystal structures and magnetic properties of full-Heusler compounds are introduced. Sections 3 and 4 demonstrate that magnetic full-Heusler compounds are promising for the TE power generation device.

2. Crystal structures and magnetic properties of full-Heusler compounds

The physical properties of full-Heusler compounds depend on their crystal structures. As shown in **Figure 2**, there are several types of crystal structures with different order degrees [4–6]. The full-Heusler compounds have four interpenetrating fcc sublattices, and each sublattice consists of the X, X', Y or Z atom. The X, X' and Y atoms are transition metals, whereas Z is a main group element. In some cases, the Y atom is a rare earth element or an alkaline earth metal.

When the X and X' atoms are of the same element, the chemical composition of the compounds is written as X_2YZ, which generally crystallises in the L2$_1$ structure. The prototype of the L2$_1$ structure is Cu_2MnAl (space group: $Fm\bar{3}m$). The Cu atoms occupy the 8*c* (1/4 1/4 1/4) site, whereas the Mn and Al atoms occupy the 4*b* (1/2 1/2 1/2) and 4*a* (0 0 0) sites, respectively. The L2$_1$ structure is a highly ordered structure of the full-Heusler compounds. Disorder among the Cu, Mn and/or Al atoms, that is, antisite defects, gives rise to different crystal structures. In a case where the Mn and Al atoms are evenly located at the 4*b* and 4*a* sites, the Cu_2MnAl becomes the B2 structure. Its prototype is CsCl (space group: $Pm\bar{3}m$). In a fully

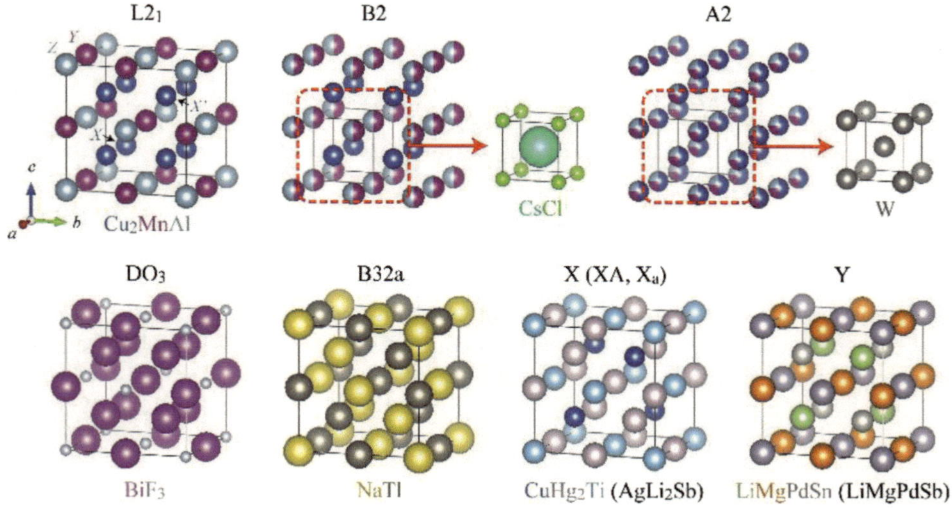

Figure 2.
Crystal structures of full-Heusler compounds. The Strukturbericht symbol and a prototype structure are written above and below each crystal structure, respectively.

disordered phase, all the atoms are randomly distributed in the 8*c*, 4*b* and 4*a* sites, thus resulting in the A2 structure. In such a structure, all the sites are equivalent, which are expressed as a bcc lattice (prototype: W, space group: $Im\bar{3}m$). There are other disordered phases, including the DO_3 and B32a structures. The former is caused by the random distribution of the *X*, *X'* and *Y* atoms at the 8*c* and 4*b* sites (prototype: BiF_3, space group: $Fm\bar{3}m$). In the B32a structure, the 8*a* (0 0 0) and 8*b* (1/2 1/2 1/2) sites are occupied by the *X*/*Y* and *X'*/*Z* atoms, respectively. The prototype is NaTl (space group: $Fd\bar{3}m$).

When the *X'* and *Y* atoms are of the same element, the chemical composition becomes XX'_2Z, which crystallises in the X (XA or X_a) structure. This structure is called the inverse Heusler phase. The prototype is $CuHg_2Ti$ (or $AgLi_2Sb$), and the space group is $F\bar{4}3\,m$. In the structure, the *X* and *Z* atoms occupy the 4*d* (3/4 3/4 3/4) and 4*a* (0 0 0) sites, respectively, and the *X'* atoms occupy the 4*b* (1/2 1/2 1/2) and 4*c* (1/4 1/4 1/4) sites.

In addition to the above ternary full-Heusler compounds, there are quaternary full-Heusler compounds, *XX'YZ*, which crystallise in the *Y* structure (prototype: LiMgPdSn or LiMgPdSb, space group: $F\bar{4}3\,m$). The *X*, *X'*, *Y* and *Z* atoms are situated at the 4*d*, 4*b*, 4*c* and 4*a* sites, respectively, occupying one of the fcc sublattices. It should be noted that the inverse Heusler and the quaternary full-Heusler phases are ordered phases, and any disorder among the constituent atoms causes a structural change; the structure changes to the B2, A2, DO_3 or B32a structure.

Earlier theoretical studies demonstrated a half-metallic nature in full-Heusler compounds [7, 8]. Since then, many studies have been dedicated to investigate the electronic and magnetic properties of ternary and quaternary full-Heusler compounds. It has been revealed that full-Heusler compounds exhibit a variety of electronic properties; they exhibit the properties of semiconductors [9–18], spin-gapless semiconductors (SGSs) [19–26], semimetals [27–29], metals [30–34] and half-metals (HMs) [32, 35–78]. Considering the magnetic properties, they have been reported to exhibit nonmagnetism [9–11, 14–18], ferromagnetism [12, 19–24, 30–33, 36–46, 48–58, 61–66, 68–78], ferrimagnetism [13, 30, 35, 47, 59, 60, 67] and antiferromagnetism [25, 26, 34]. The full-Heusler, inverse Heusler and quaternary Heusler compounds obey the Slater-Pauling rule [79–81]: the total spin magnetic moment per unit cell scales with the total number of valence electrons in the unit cell.

3. Thermoelectric properties of half-metallic full-Heusler compounds

In this section, we present some of the theoretical and experimental studies on the TE properties of half-metallic full-Heusler compounds. The TE properties can be calculated on the basis of the Boltzmann transport equations [82–84]. Using the electronic energy-wavenumber dispersion curve of the i-th band $\varepsilon_i(\boldsymbol{k})$, the tensors of the Seebeck coefficient, $\boldsymbol{S}(T)$, electrical conductivity, $\boldsymbol{\sigma}(T)$, and carrier thermal conductivity, $\boldsymbol{\kappa}_{\mathrm{e}}(T)$, can be expressed as:

$$\boldsymbol{S}(T) = -\frac{1}{|e|T}\frac{\int_{-\infty}^{+\infty}\tilde{\boldsymbol{\sigma}}(\varepsilon,T)(\varepsilon-\varepsilon_{\mathrm{F}})\left(-\frac{\partial f_{\mathrm{FD}}(\varepsilon,T)}{\partial\varepsilon}\right)\mathrm{d}\varepsilon}{\boldsymbol{\sigma}(T)}, \tag{3}$$

$$\boldsymbol{\sigma}(T) = \int_{-\infty}^{+\infty}\tilde{\boldsymbol{\sigma}}(\varepsilon,T)\left(-\frac{\partial f_{\mathrm{FD}}(\varepsilon,T)}{\partial\varepsilon}\right)\mathrm{d}\varepsilon, \tag{4}$$

$$\boldsymbol{\kappa}_{\mathrm{e}}(T) = -\frac{1}{e^2T}\int_{-\infty}^{+\infty}\tilde{\boldsymbol{\sigma}}(\varepsilon,T)(\varepsilon-\varepsilon_{\mathrm{F}})^2\left(-\frac{\partial f_{\mathrm{FD}}(\varepsilon,T)}{\partial\varepsilon}\right)\mathrm{d}\varepsilon - \frac{1}{e^2T}\frac{\left\{\int_{-\infty}^{+\infty}\tilde{\boldsymbol{\sigma}}(\varepsilon,T)(\varepsilon-\varepsilon_{\mathrm{F}})\left(-\frac{\partial f_{\mathrm{FD}}(\varepsilon,T)}{\partial\varepsilon}\right)\mathrm{d}\varepsilon\right\}^2}{\boldsymbol{\sigma}(T)}, \tag{5}$$

$$\tilde{\sigma}_{\alpha\beta}(\varepsilon,T) \equiv \frac{1}{N_k}\sum_{i,k}\frac{e^2\tau(k,T)}{\hbar^2}\frac{\partial\varepsilon_i(k)}{\partial k_\alpha}\frac{\partial\varepsilon_i(k)}{\partial k_\beta}\delta(\varepsilon-\varepsilon_i(k)), (\alpha,\beta=x,y,z), \tag{6}$$

where e, ε, ε_{F}, $f_{\mathrm{FD}}(\varepsilon,T)$, N_k, $\tau(k,T)$, and $\tilde{\boldsymbol{\sigma}}(\varepsilon,T)$ are the elementary charge, electron energy, Fermi level, Fermi-Dirac distribution function, total number of the k-points, relaxation time, Dirac constant and conductance spectrum tensor, respectively. It is difficult to calculate the relaxation time; hence, the calculation of TE properties generally gives $\boldsymbol{S}(T)$, $\boldsymbol{\sigma}(T)/\tau$ and $\boldsymbol{\kappa}_{\mathrm{e}}(T)/\tau$ [84]. In context to magnetic materials, the electronic states of the majority and minority spin electrons are considered. Assuming that τ for the majority and minority spin electrons is the same, the total S for the magnetic materials, $S_{\mathrm{tot}}(T)$, is calculated by

$$S_{\mathrm{tot}}(T) = \frac{S_\uparrow(T)\sigma_\uparrow(T)/\tau + S_\downarrow(T)\sigma_\downarrow(T)/\tau}{\sigma_\uparrow(T)/\tau+\sigma_\downarrow(T)/\tau} = \frac{S_\uparrow(T)\sigma_\uparrow(T) + S_\downarrow(T)\sigma_\downarrow(T)}{\sigma_\uparrow(T)+\sigma_\downarrow(T)}, \tag{7}$$

where S and σ with the up- and down-arrow subscripts those evaluated from the electronic states of the majority and minority spin electrons, respectively.

Figure 3(a) and **(b)** shows the temperature dependence of the calculated S_{tot} for ternary and quaternary half-metallic full-Heusler compounds, respectively. To calculate the electronic band, the full-potential linearised augmented plane wave (FLAPW) method was employed, adopting the local spin density approximation (LSDA) or the generalised gradient approximation in the Perdew-Burke-Ernzerhof parametrisation (PBE-GGA) as the local exchange-correlation potential. As seen in the figure, the negative and positive S_{tot} are presented, indicating that both n-type and p-type materials can be obtained from half-metallic full-Heusler compounds. The S_{tot} is observed to increase with increasing temperature for almost all the compounds, which is the typical behaviour of metal. Furthermore, the S_{tot} is observed to attain values as high as several tens of μV/K. These values are lower

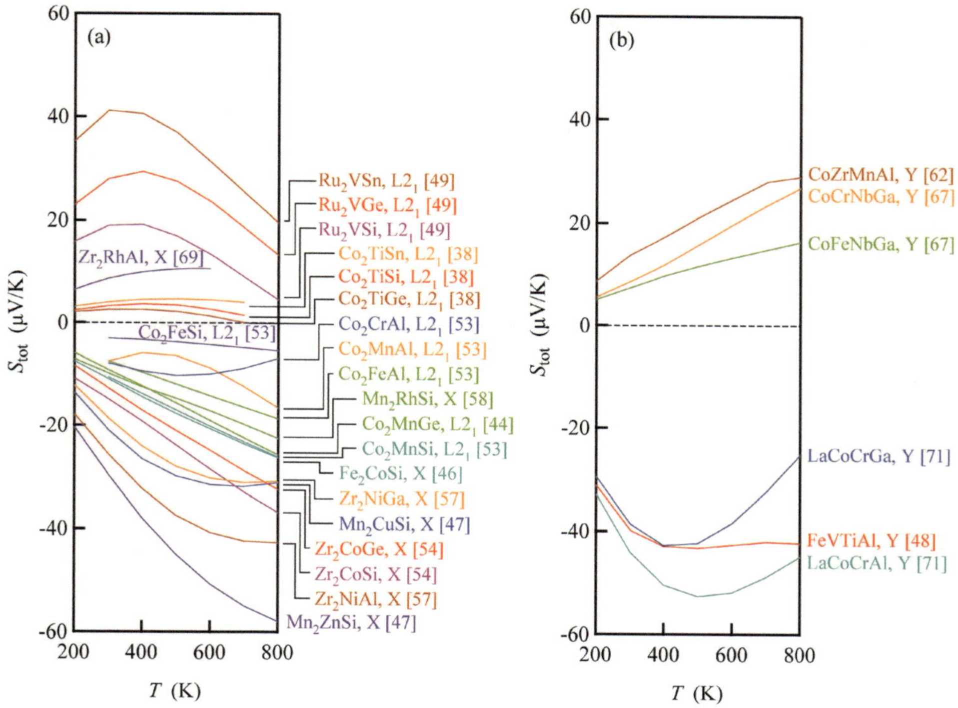

Figure 3.
Temperature dependence of the calculated S_{tot} for half-metallic full-Heusler compounds. Their crystal structures are also shown. The calculation of the electronic structure was performed using the full-potential linearised augmented plane wave (FLAPW) method with local spin density approximation (LSDA) or generalised gradient approximation in the Perdew-Burke-Ernzerhof parametrisation (PBE-GGA).

than those of TE semiconductors but higher than those of common metals, demonstrating the potential of half-metallic full-Heusler compounds as high-temperature TE materials.

The temperature dependence of S for several half-metallic Co-based full-Heusler compounds was determined by Balke et al. [37] and Hayashi et al. [53]. For the measurements, the S_{tot} values for the compounds were obtained. Hereafter, we use S to represent S_{tot}. As shown in **Figure 4(a)–(c)**, the Co-based full-Heusler compounds exhibit negative S in the order of several tens of μV/K. For metals, the sign of S is well explained by Mott's formula [85]:

$$S \propto -\frac{1}{\mathrm{DOS}(\varepsilon_F)} \frac{\mathrm{d\,DOS}(\varepsilon)}{\mathrm{d}\varepsilon}\bigg|_{\varepsilon=\varepsilon_F}, \tag{8}$$

where DOS is the electronic density of states. Adopting Eq. (8) for the partial DOS of the *sp*-electrons and *d*-electrons of Co_2MnSi, it was obtained that in half-metallic full-Heusler compounds, the itinerant *sp*-electrons contribute more to S than the localised *d*-electrons [53]. In **Figure 4**, Co_2TiAl is shown to exhibit the highest $|S|$ of $|-56|$ μV/K at 350 K among other compounds. It is observed that Co_2TiSi, Co_2TiGe and Co_2TiSn exhibit a characteristic temperature dependence of $|S|$; the value of $|S|$ increases with increasing temperature and becomes constant at temperatures above 350 K. This characteristic behaviour is further discussed later in this section.

The half-metallic full-Heusler compounds are predicted to have high electrical conductivity σ owing to their metallic properties; hence, they are considered to be superior to the semiconductors. **Figure 5(a)** shows the temperature dependence of

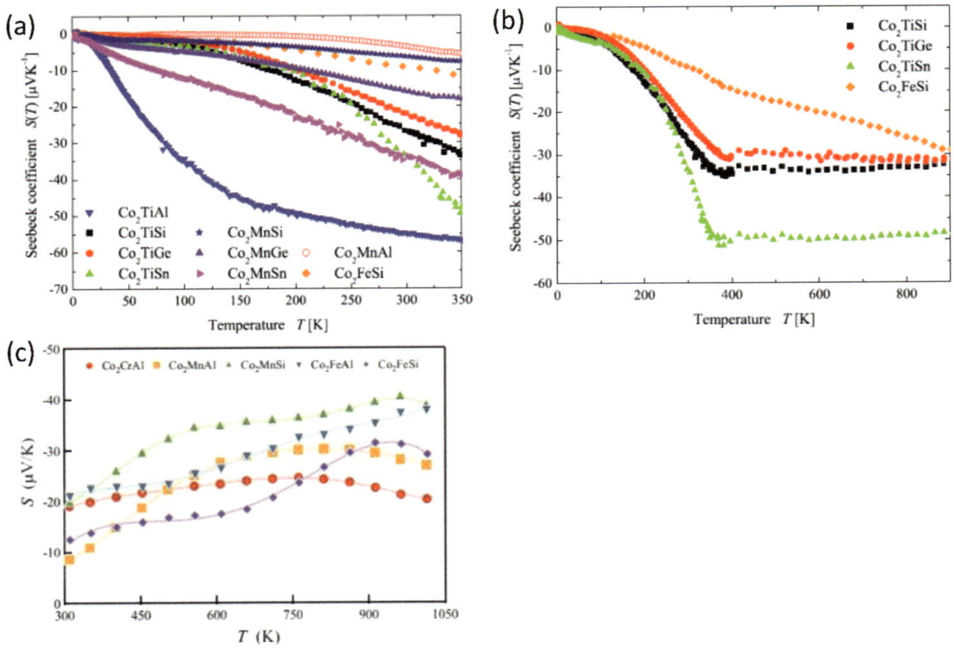

Figure 4.
Temperature dependence of the measured S of several Co-based full-Heusler compounds. ((a) and (b) Reprinted from [37]. Copyright 2010, with permission from Elsevier. (c) Reprinted from [53]. Copyright 2017, with permission from Springer).

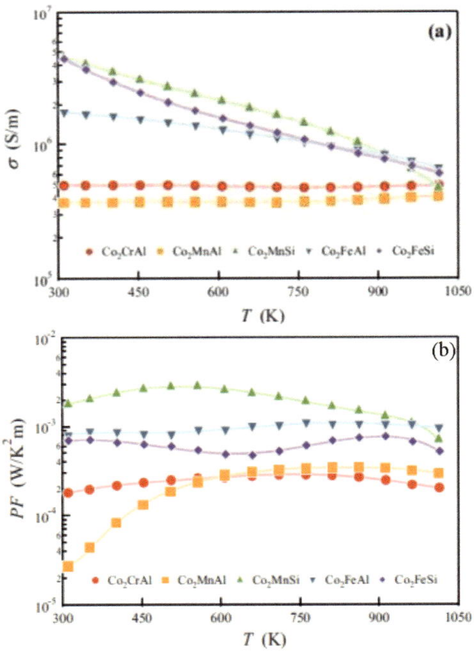

Figure 5.
(a) Measured σ and (b) PF of several Co-based full-Heusler compounds as a function of temperature. (Reprinted from [53]. Copyright 2017, with permission from Springer).

the measured σ for several Co-based full-Heusler compounds [53]. The σ values of the compounds are observed to be high, ranging from 10^5 to 10^7 S/m. Among all the compounds, Co_2MnSi exhibits the highest σ in the whole temperature range. The σ

value of Co_2MnSi decreases from 4.6×10^6 S/m at 300 K to 4.7×10^5 S/m at 1000 K. This is a typical electrical conductivity-temperature relation in metals. From the S and σ values (shown in **Figures 4(c)** and **5(a)**, respectively), the PF was calculated and plotted in **Figure 5(b)** [53]. Owing to the high S and high σ, Co_2MnSi exhibits the highest PF (2.9×10^{-3} W/K^{2}m at 500 K) among other compounds, which is comparable to that of a Bi_2Te_3-based material [86]. Since Co_2MnSi exhibits a negative S, it could be a potential n-type TE material. Thus, to develop a TE device using full-Heusler compounds, a p-type counterpart to Co_2MnSi is needed. For this purpose, Li et al. [60, 78] prepared a half-metallic Mn_2VAl compound and measured its TE properties. Although Mn_2VAl is a p-type material showing positive S, its highest PF (2.84×10^{-4} W/K^{2}m at 767 K [78]) is lower than that of Co_2MnSi. Thus, there is a need to explore more p-type half-metallic full-Heusler compounds with high PF.

Here, the temperature dependence of S for the various full-Heusler compounds is discussed. Comparing the calculated S values for Co_2TiSi, Co_2TiGe and Co_2TiSn (**Figure 3(a)**) with the measured values (**Figure 5(b)**), it is obtained that not only the temperature dependence but also the sign of the S values are different. As mentioned earlier, the measured S value is almost constant at temperatures above 350 K; however, the calculated values do not display such relation. To explain this difference, Barth et al. [38] considered the difference in the electronic structure of the ferromagnetic (FM) state and nonmagnetic (NM) states. They obtained that the FM-NM phase transition occurs around 350 K for Co_2TiSi, Co_2TiGe and Co_2TiSn [38]. Using the temperature dependence of S for the FM and NM states, $S_{FM}(T)$ and $S_{NM}(T)$, and that of the normalised magnetisation calculated by using the molecular field theory, $M(T)$, a modified S value, S_{FM+NM}, can be calculated according to the formula [38]:

$$S_{FM+NM}(T) = \frac{S_{FM}(T)\sigma_{FM}(T)M(T) + S_{NM}(T)\sigma_{NM}(T)\{1 - M(T)\}}{\sigma_{FM+NM}(T)}, \tag{9}$$

where σ_{FM+NM} is the modified electrical conductivity of a mixture of FM and NM states weighted by using $M(T)$. Although the above consideration is plausible, the calculated S_{FM+NM} values for Co_2TiSi, Co_2TiGe and Co_2TiSn (**Figure 6**) do not coincide with the measured values. The inconsistency between the S_{FM+NM} values and the measured ones is also observed in the case of Co_2CrAl, Co_2MnAl, Co_2MnSi, Co_2FeAl and Co_2FeSi [53].

It is suggested that the constant S value in the NM state for Co_2TiSi, Co_2TiGe and Co_2TiSn (**Figure 4(b)**) is governed by the relaxation time rather than by the electronic structure [38]. The S value is calculated by using Eq. (1), where both the numerator and denominator of the fraction are functions of relaxation time $\tau(k, T)$; τ is included in both numerator and denominator of the fraction through $\tilde{\boldsymbol{\sigma}}(\varepsilon, T)$ described in Eq. (6). However, in the calculation, the τ in the numerator and denominator cancels each other. In addition, the total S is calculated assuming that τ for the majority and minority spin electrons is the same (Eq. (7)). The neglected τ in Eqs. (1) and (7) could be a reason for the difference in the temperature dependence of the calculated and measured S. Another possible reason for this discrepancy is the method employed in calculating the electronic structure. The calculation results shown in **Figures 3** and **6** are based on the LSDA or PBE-GGA. The use of the onsite Hubbard interaction in combination with PBE-GGA, namely, PBE + U or GGA + U [51, 55, 70, 73], and the Tran-Blaha modified Becke-Johnson (TB-MBJ) [64, 73] gives electronic structures different from that obtained using the LSDA or PBE-GGA, which may lead to a temperature dependence of S well-fitted to the measured one.

Also, defect and/or disorder in full-Heusler compounds affect the temperature dependence, as well as the sign of S, which could be another reason for the

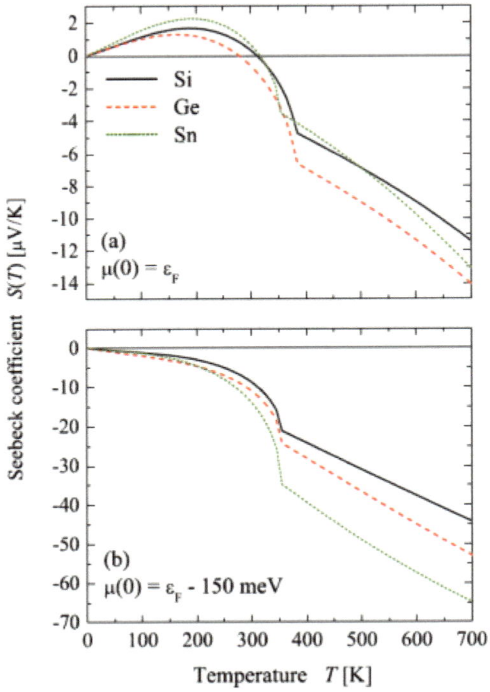

Figure 6.
Temperature dependence of the calculated S *of Co_2TiZ (Z = Si, Ge, Sn) considering the FM-NM phase transition. In the calculation, the chemical potential at* T = 0, $\mu(0)$, *was set to (a) ε_F and (b) 150 meV below ε_F. (Reprinted from [38]. Copyright 2010, with permission from American Physical Society).*

discrepancy in the temperature dependence of *S*. The structure model used for the calculation in **Figures 3** and **6** is the $L2_1$, X or Y structure, which is highly ordered phases, devoid of any defect, for the ternary and quaternary full-Heusler compounds. Popescu et al. [52] investigated the effect of several defects on the temperature dependence of *S* for Co_2TiZ (*Z* = Si, Ge, Sn) in the FM state. As shown in **Figure 7**, off-stoichiometric defects, such as Co vacancy and the substitution of excess atoms at a particular site, change the sign of *S*.

The effect of structural disorder on *S* for Co_2CrAl, Co_2MnAl, Co_2MnSi, Co_2FeAl and Co_2FeSi has been obtained, as shown in **Figure 8** [53]. The figure compares the calculated S_{FM+NM} with the measured *S*. It is observed that the measured values of *S* are individually higher than the calculated value (S_{FM+NM}). Considering the crystal structure, Co_2CrAl, Co_2MnAl, Co_2MnSi, Co_2FeAl and Co_2FeSi are not in the fully ordered $L2_1$ structure; most of them crystallise in the disordered B2 and/or A2 structures. This result implies that the B2 and/or A2 structures exhibit higher *S* than the $L2_1$ structure. Recently, Li et al. [78] investigated the effect of structural disorder on the value of *S* for half-metallic Mn_2VAl compounds by varying the B2 order degree. **Figure 9(a)** shows the measured *S* values for Mn_2VAl with the B2 order degree of 27 and 66%. The *S* values for the structure having 66% B2 order degree are observed to be higher than those for 27% B2 order degree in the entire measurement temperature range. In addition, it is observed that the *S* value increases with increasing the B2 order degree (**Figure 9(b)**). The increase in the B2 order degree means an increase in the disorder between the V and Al atoms, that is, a decrease in the $L2_1$ order degree. To understand the reason for the difference in *S* between the $L2_1$ and B2 structures, the DOS of Mn_2VAl with the $L2_1$ and B2 structures was calculated by using the Korringa-Kohn-Rostoker method. It was obtained that the B2 structure exhibits a steeper DOS of the majority-spin *sp*-electrons than the $L2_1$

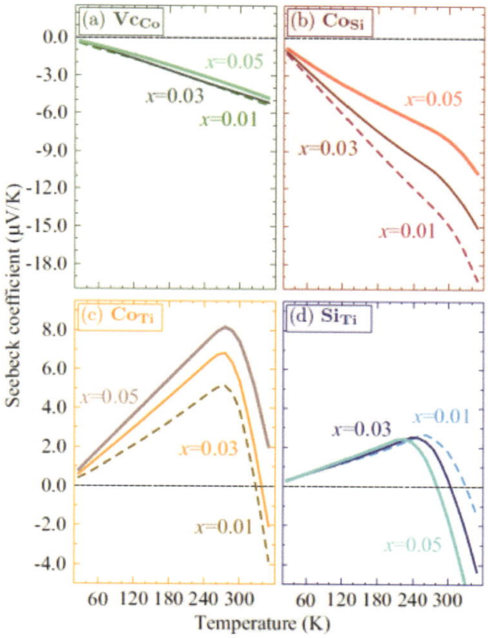

Figure 7.
Change in calculated S for Co_2TiSi with several off-stoichiometric defects such as (a) Co vacancy (V_{Co}), (b) excess Co atoms at the Si site (Co_{Si}), (c) excess Co atoms at the Ti site (Co_{Ti}) and (d) excess Si atoms at the Ti site (Si_{Ti}). (Reprinted from [52]. Copyright 2017, with permission from American Physical Society).

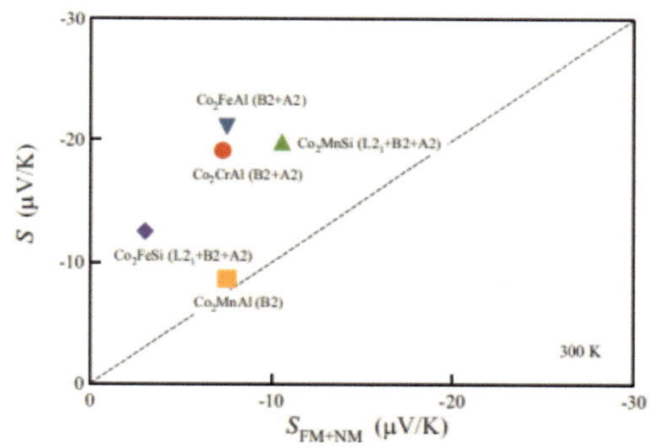

Figure 8.
Comparison between the calculated $S_{FM + NM}$ and the measured S at 300 K for several Co-based full-Heusler compounds. (Reprinted from [53]. Copyright 2017, with permission from Springer).

structure, which is considered as the main reason for the higher *S* of the B2 structure than that of the $L2_1$ structure. Further increase in the B2 order degree is expected to yield a higher *S* for Mn_2VAl. The modulation of the order degree can be a key strategy to enhance the *S* value of the half-metallic full-Heusler compounds; the disorder in Co_2CrAl, Co_2MnAl, Co_2MnSi, Co_2FeAl, Co_2FeSi and Mn_2VAl gives rise to the higher *S*. To establish this strategy, the effects of the order degree, not only on *S* but also on σ, should be investigated for several half-metallic full-Heusler compounds.

Considering the TE performance of the half-metallic full-Heusler compounds, not only PF but also zT are important. To evaluate the zT of Co_2MnSi, we obtained the temperature dependence of the total thermal conductivity, κ_{tot} (**Figure 10(a)**). Similar to the case of common metals, a high κ_{tot} was obtained. It decreases with increasing temperature from 79 W/Km at 300 K to 21 W/Km at 1000 K.

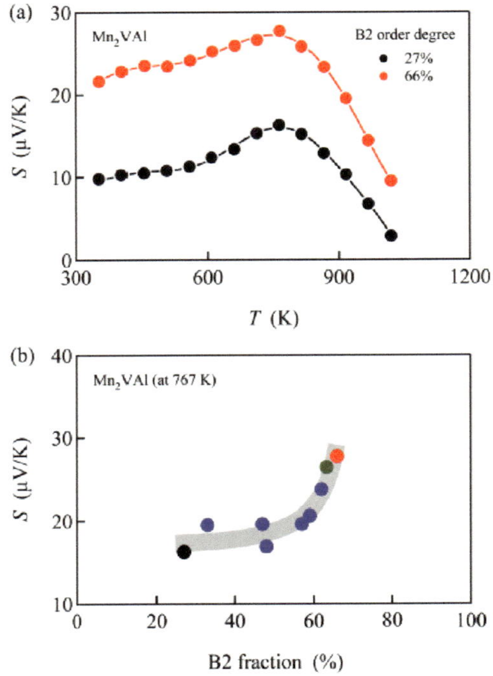

Figure 9.
(a) Temperature dependence of the measured S of Mn_2VAl with the B2 order degree of 27 and 66%. (b) Measured S values of Mn_2VAl at 767 K plotted against the B2 order degree. (Reprinted from [78]. Copyright 2020, with permission from IOP Publishing).

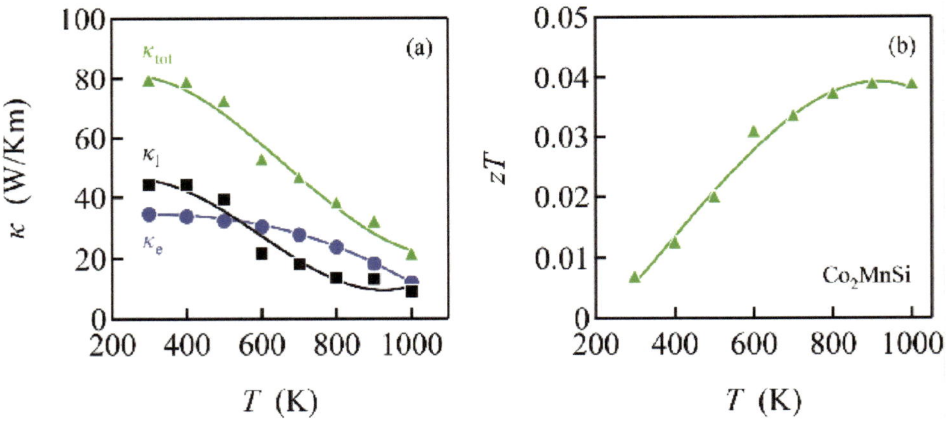

Figure 10.
Temperature dependence of (a) measured κ_{tot}, κ_e and κ_l and (b) evaluated zT of Co_2MnSi.

Figure 10(b) shows the temperature dependence of zT for Co_2MnSi calculated using the PF value (**Figure 5(b)**) and the κ_{tot} value (**Figure 10(a)**). Due to the high κ_{tot}, the maximum zT value, zT_{max}, of Co_2MnSi is 0.039, which is obtained at temperatures above 900 K. Although this zT_{max} value is far below the standard level of $zT = 1$, it is higher than that of Co_2TiSn (0.033 at 370–400 K) [38] and those of semi-metallic full-Heusler compounds (0.0052 at 300 K for Ru_2NbAl [28] and 0.0027 at 300 K for $Ru_2VAl_{0.25}Ga_{0.75}$ [29]).

It should be noted that the κ_{tot} of Co_2MnSi is not equal to the carrier thermal conductivity, κ_e. The κ_e value can be calculated by using the Wiedemann-Franz law, $\kappa_e = L\sigma T$, where L is the Lorentz number. Evaluating the L value on the basis of the single parabolic band model [87] and using the measured σ value (**Figure 5(a)**), the κ_e value of Co_2MnSi was calculated and plotted in **Figure 10(a)**. It can be observed from the figure that κ_e is only half as high as κ_{tot}. The rest is attributed to the lattice thermal conductivity, κ_l ($=\kappa_{tot} - \kappa_e$), as shown in **Figure 10(a)**, which amounts to a half of the κ_{tot}. This is contrary to the case of common metals where the κ_{tot} is mainly dominated by κ_e [88]. The non-negligible κ_l suggests that, for the theoretical evaluation of zT of the half-metallic full-Heusler compounds, the contribution of κ_l should not be ignored. Experimentally, the high contribution of κ_l to κ_{tot} indicates that the κ_{tot} of half-metallic full-Heusler compounds could be reduced by decreasing the κ_l.

4. Future prospects of magnetic full-Heusler compounds as potential thermoelectric materials

In this section, we introduce other full-Heusler compounds to demonstrate the potentials of magnetic full-Heusler compounds in TE applications. First, we consider the full-Heusler SGSs as an example. Schematic illustrations of the DOS of SGSs and HMs are shown in **Figure 11**. The DOS of SGSs has an open band gap in one spin electron and a closed gap in the other. Since the Fermi level ε_F is located just at the closed gap, the electron or hole concentration in SGSs is expected to be less than that in HMs. One of the investigated SGSs is the full-Heusler Mn_2CoAl, which crystallises in the X structure (the inverse Heusler phase). The variation of its σ, S and carrier concentration, n, with temperature is shown in **Figure 12**, as determined by Ouardi et al. [19]. It can be observed that the σ and n vary slightly with the temperature, which is attributed to the typical behaviour of gapless semiconductors [89]. In addition, the S value is nearly equal to 0 μV/K. The reduced Seebeck effect indicates the occurrence of electron and hole compensation, which is the evidence that ε_F is at the top of the valence states and at the bottom of the conduction states.

Owing to the nearly zero S values, Mn_2CoAl cannot be used as a TE material; however, there is a possibility of achieving high $|S|$ in Mn_2CoAl by tuning the position of ε_F. The position of ε_F can be varied via partial substitution, which increases the hole or electron carrier concentration in Mn_2CoAl. We calculated the S value for the partially substituted Mn_2CoAl, as shown in **Figure 13(a)**. The calculation was based on a rigid band model; thus, the electronic structure of the partially substituted Mn_2CoAl is assumed to be the same as that of Mn_2CoAl. In the figure, the horizontal axis is μ-ε_F, where μ and ε_F are the chemical potential (i.e., the Fermi

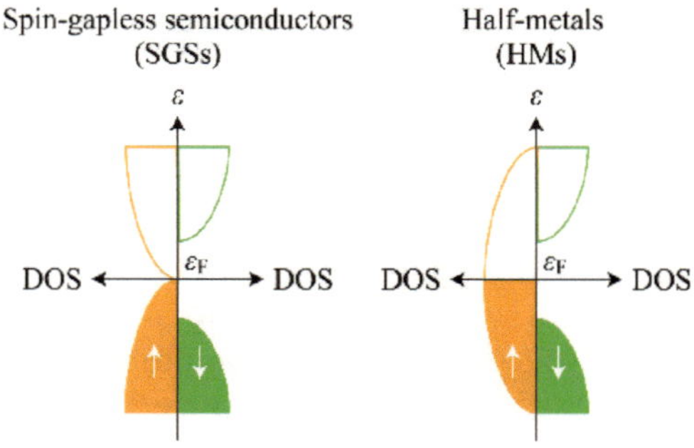

Figure 11.
Schematic illustration of DOS for spin-gapless semiconductors (SGSs) and half-metals (HMs).

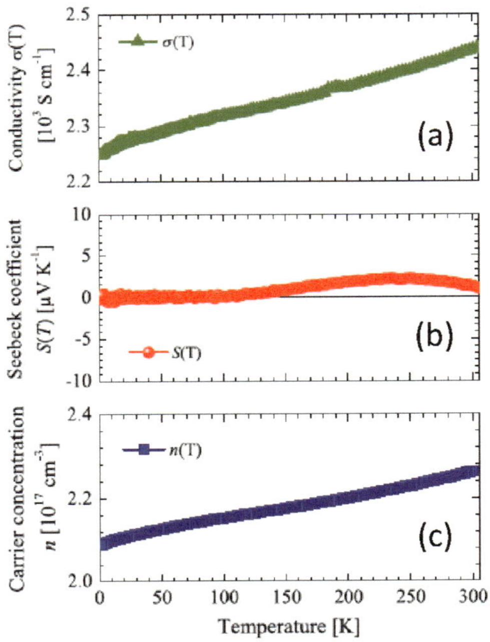

Figure 12.
Temperature dependence of the measured (a) σ, (b) S and (c) n of Mn_2CoAl. (Reprinted from [19]. Copyright 2020, with permission from American Physical Society).

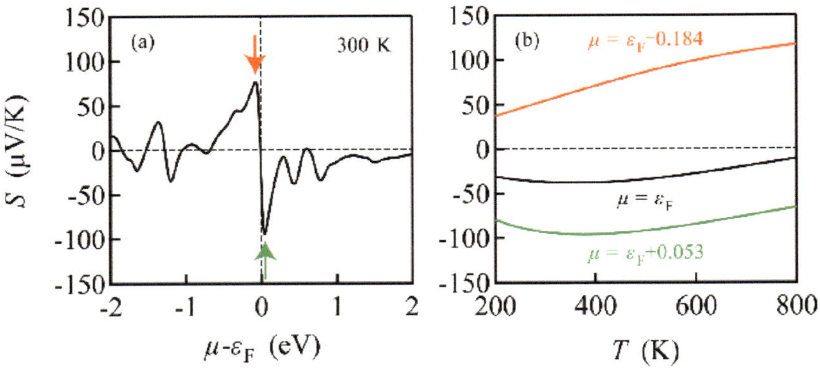

Figure 13.
(a) Calculated S value at 300 K for the partially substituted Mn_2CoAl plotted as a function of μ-ε_F, where μ and ε_F are the Fermi levels of partially substituted Mn_2CoAl and that of Mn_2CoAl, respectively. The highest |S| values are obtained at $\mu = \varepsilon_F - 0.184$ and at $\mu = \varepsilon_F + 0.053$, as denoted by orange and green arrows, respectively. (b) Temperature dependence of S at $\mu = \varepsilon_F$, $\mu = \varepsilon_F - 0.184$ and $\mu = \varepsilon_F + 0.053$.

level) of the partially substituted Mn_2CoAl and the Fermi level of Mn_2CoAl, respectively. A negative/positive μ-ε_F value means an increase in the hole/electron carrier concentration. Although the value of S at $\mu = \varepsilon_F$, corresponding to the case of Mn_2CoAl, is small, it is large at $\mu = \varepsilon_F - 0.184$ and at $\mu = \varepsilon_F + 0.053$ (pointed by orange and green arrows, respectively). The temperature dependences of S at $\mu = \varepsilon_F$, $\mu = \varepsilon_F - 0.184$ and $\mu = \varepsilon_F + 0.053$ are shown in **Figure 13(b)**, which again demonstrates that high $|S|$ values can be achieved for both p-type and n-type regions. These calculation results prove the full-Heusler SGSs as potential materials for TE applications.

To achieve high |*S*| values for Mn_2CoAl, it is important to retain its SGS characteristic. Galanakis et al. [90] theoretically investigated the effects of structural disorder on the electronic structure of Mn_2CoAl. It was obtained that the SGS characteristic is not conserved in the presence of Mn-Co, Mn-Al and Co-Al antisite defects. Instead of the closed gap, low DOS intensity emerges in the electronic structure of the majority spin electrons around ε_F, indicating that the disorder induces half-metallic characteristics in Mn_2CoAl. Also, Xu et al. [91] reported that an as-prepared Mn_2CoAl compound is non-stoichiometric and contains the Mn-Co antisite defect. In a case where Mn_2CoAl is not an SGS, the |*S*| cannot be increased via partial substitutions.

Other examples considered here are the full-Heusler compounds having low values of κ_l. **Figure 14** shows a flower-like microstructure of $Co_2Dy_{0.5}Mn_{0.5}Sn$ observed by Schwall et al. [43]. Although the chemical composition of $Co_2Dy_{0.5}Mn_{0.5}Sn$ coincides with that of the full-Heusler phase, $Co_2Dy_{0.5}Mn_{0.5}Sn$ is not in a single phase. It consists of two major phases: half-metallic Co_2MnSn and ferromagnetic $Co_8Dy_3Sn_4$ phases. This phase separation is induced by rapid cooling from the liquid phase. Consequently, the κ_l value of $Co_2Dy_{0.5}Mn_{0.5}Sn$ is lower than those of Co_2MnSn and $Co_8Dy_3Sn_4$.

He et al. [9] theoretically discovered a new class of stable nonmagnetic full-Heusler semiconductors with high PF and ultralow κ_l, attributed to atomic rattling. The compounds contain alkaline earth elements (Ba, Sr or Ca) in the *X* sublattice and noble metals (Au or Hg) and main group elements (Sn, Pb, As or Sb) in the *Y* and *Z* sublattices, respectively. The κ_l value of Ba_2AuBi and Ba_2HgPb was obtained to be lower than 0.5 W/Km at 300 K. At higher temperatures, it was close to the theoretical minimum, that is, the amorphous limit of 0.27 W/Km [92]. Park et al. [16] further examined the TE properties of Ba_2BiAu. They predicted that considerably high zT of ~5 can be achieved at 800 K.

Finally, there are many ternary and quaternary full-Heusler compounds yet to be explored. Among the full-Heusler compounds, nonmagnetic Fe_2VAl-based compounds have been intensively investigated as one of the potential TE semiconductors [93]. Despite the long historical investigation, Hinterleitner et al. [94] discovered quite recently that a metastable $Fe_2V_{0.8}W_{0.2}Al$ thin film exhibits extremely high zT of ~6 at 350 K as a result of its high *S*. The crystal structure of the thin film is reported to be the disordered A2 structure, which could be the reason for its high *S*, as in the cases of several half-metallic Co-based and Mn-based full-Heusler compounds [53, 78]. If the disorder in structure contributes to the high *S*, then the strategy of enhancing zT by controlling structural disorder would be applicable to the other full-Heusler compounds. Herewith, more conventional and novel findings on the full-Heusler compounds can be achieved.

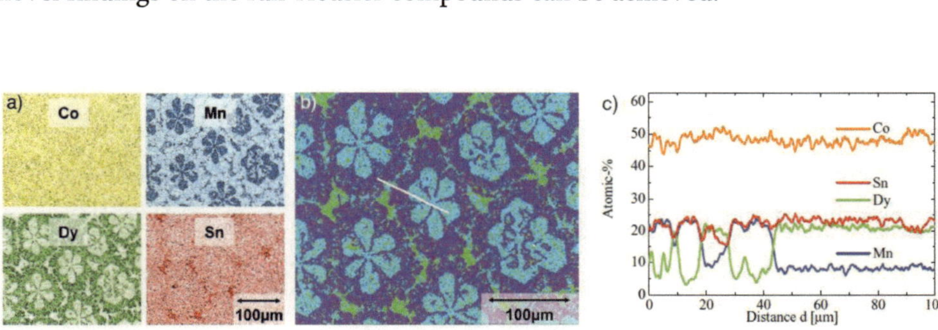

Figure 14.
Flower-like microstructure of $Co_2Dy_{0.5}Mn_{0.5}Sn$. (a) Elemental mappings, (b) combined image of elemental mappings shown in (a). (c) Line scan along the line indicated in (b). (Reprinted from [43]. Copyright 2012, with permission from WILEY-VCH).

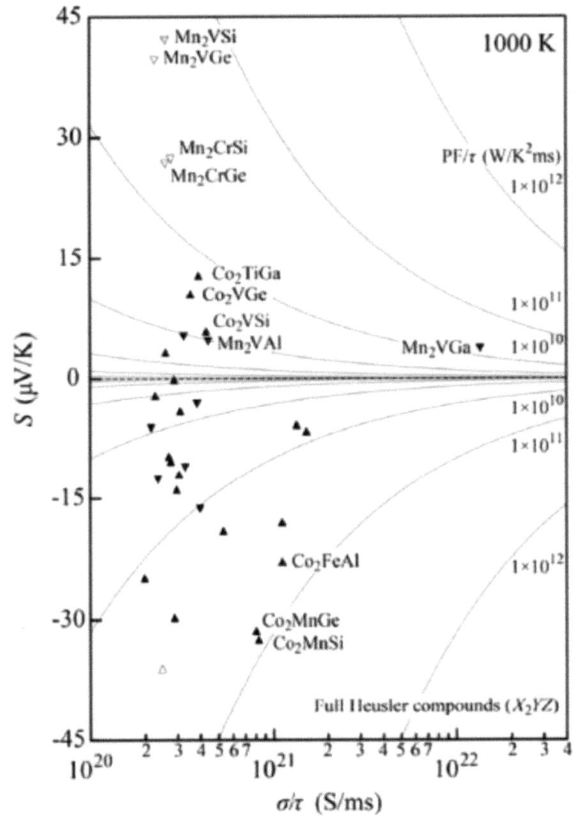

Figure 15.
Calculated S versus σ/τ at 1000 K for several Co-based and Mn-based full-Heusler compounds. The grey curves indicate PF/τ. (Reprinted from [60]. Copyright 2018, with permission from Elsevier).

To explore the potentials of the full-Heusler compounds, theoretical studies are vital to minimise the experimental tasks. **Figure 15** exhibits a plot of S versus σ/τ at 1000 K for several Co-based and Mn-based full-Heusler compounds, as calculated by Li et al. [60]. Furthermore, recent advancements in machine learning dispel the difficulty in searching novel full-Heusler compounds [95, 96]. Combining such calculations with experiments, we can effectively discover magnetic full-Heusler compounds with much higher TE efficiency, which promises the realisation of high-efficiency TE power generation devices.

Acknowledgements

We greatly acknowledge the financial supports from the Thermal and Electric Energy Technology Foundation and from the Tsinghua-Tohoku Collaborative Research Fund.

Conflict of interest

We declare that there is no conflict of interest.

Nomenclatures

$\eta_{\max}$ [−]	maximum TE efficiency
zT [−]	dimensionless figure-of-merit
T [K]	absolute temperature
S [V/K]	Seebeck coefficient
σ [S/m]	electrical conductivity
κ [W/Km]	thermal conductivity
PF [W/K^{2}m]	power factor
ε_i [eV]	electronic energy–wavenumber dispersion curve of the i-th band
$\boldsymbol{k}$ [m^{-1}]	wavenumber
$\boldsymbol{S}$ [V/K]	Seebeck coefficient tensor
$\boldsymbol{\sigma}$ [S/m]	electrical conductivity tensor
$\boldsymbol{\kappa}_{\mathrm{e}}$ [W/Km]	carrier thermal conductivity tensor
$e = 1.60217662 \times 10^{-19}$ C	elementary charge
ε [eV]	electron energy
ε_{F} [eV]	Fermi level
f_{FD} [−]	Fermi-Dirac distribution function
N_k [−]	total number of k-points
τ [1/s]	relaxation time
$= 6.582119569 \times 10^{-16}$ eVs	Dirac constant
$\tilde{\boldsymbol{\sigma}}$ [S/m]	conductance spectrum tensor
DOS [states/eV]	density of states
M [−]	normalised magnetisation calculated by using molecular field theory
μ [eV]	chemical potential
κ_{e} [W/Km]	carrier thermal conductivity
L [WΩ/K^2]	Lorentz number
κ_{l} [W/Km]	lattice thermal conductivity
n [1/m^3]	carrier concentration

Author details

Kei Hayashi*, Hezhang Li, Mao Eguchi, Yoshimi Nagashima and Yuzuru Miyazaki
Department of Applied Physics, Graduate School of Engineering, Tohoku University, Sendai, Japan

*Address all correspondence to: hayashik@crystal.apph.tohoku.ac.jp

References

[1] Rowe DM, editor. Thermoelectrics Handbook: Macro to Nano. Boca Raton, FL: CRC Press; 2005. p. 1022

[2] Uher C. Materials Aspect of Thermoelectricity. Boca Raton, FL: CRC Press; 2016. p. 610

[3] Kurosaki K, Takagiwa Y, Shi X, editors. Thermoelectric Materials: Principles and Concepts for Enhanced Properties. Berlin: De Gruyter; 2020. p. 280

[4] Heusler F, Starck W, Haupt E. Magnetisch-chemische Studien. Verhandlungen der Deutschen Physikalischen Gesellschaft. 1903;**5**: 219-232 (in German)

[5] Webster P. Heusler alloys. Contemporary Physics. 1969;**10**:559-577. DOI: 10.1080/00107516908204800

[6] Webster PJ, Ziebeck KRA. Heusler alloys. In: Wijn HRJ, editor. Landolt-Börnstein New Series Group III. Vol. 19C. Berlin: Springer; 1988. pp. 75-185

[7] Kübler J, William AR, Sommers CB. Formation and coupling of magnetic moments in Heusler alloys. Physical Review B. 1983;**28**:1745-1755. DOI: 10.1103/PhysRevB.28.1745

[8] Ishida S, Fujii S, Kashiwagi S, Asano S. Search for half-metallic compounds in Co_2MnZ (Z=IIIb, IVb, Vb element). Journal of the Physical Society of Japan. 1995;**64**:2152-2157. DOI: 10.1143/JPSJ.64.2152

[9] He J, Amsler M, Xia Y, Naghavi SS, Hegde VI, Hao S, et al. Ultralow thermal conductivity in full Heusler semiconductors. Physical Review Letters. 2016;**117**:046602. DOI: 10.1103/PhysRevLett.117.046602

[10] Li J, Li J, Zhang Q, Zhang Z, Yang G, Ma H, et al. The thermoelectric properties of predicted semiconducting Ti_2CrGe and Ti_2CrSn: A first principles study. Computational Materials Science. 2016;**125**:183-187. DOI: 10.1016/j.commatsci.2016.08.048

[11] Li J, Yang G, Yang Y, Ma H, Zhang Q, Zhang Z, et al. Electronic and thermoelectric properties of nonmagnetic inverse Heusler semiconductors Sc_2FeSi and Sc_2FeGe. Journal of Magnetism and Magnetic Materials. 2017;**442**:371-376. DOI: 10.1016/j.jmmm.2017.07.010

[12] Yousuf S, Gupta DC. Insight into electronic, mechanical and transport properties of quaternary CoVTiAl: Spin-polarized DFT + U approach. Materials Science and Engineering B. 2017;**221**: 73-79. DOI: 10.1016/j.mseb.2017.04.004

[13] Zhang Q, Ma H, Yang G, Yang Y, Li J, Li C, et al. Prediction of fully compensated ferrimagnetic and nonmagnetic semiconductors with promising thermoelectric properties through the Mo substitution of Cr for Ti_2CrZ (Z=Ge, Sn) Heusler alloys. Intermetallics. 2018;**96**:72-78. DOI: 10.1016/j.intermet.2018.03.001

[14] Yan Y, Yang J, Li J, Wang Y, Ren W. High thermoelectric properties in full-Heusler X_2YZ alloys (X = Ca, Sr, and Ba; Y = Au and Hg; Z = Sn, Pb, As, Sb, and Bi). Journal of Physics D: Applied Physics. 2019;**52**:495303. DOI: 10.1088/1361-6463/ab40e6

[15] Matougui M, Bouadjemi B, Houari M, Haid S, Lantri T, Zitouni A, et al. Rattling Heusler semiconductors' thermoelectric properties: First-principles prediction. Chinese Journal of Physics. 2019;**57**:195-210. DOI: 10.1016/j.cjph.2018.11.015

[16] Park J, Xia Y, Ozoliņš V. High thermoelectric power factor and efficiency from a highly dispersive band in Ba_2BiAu. Physical Review Applied.

2019;**11**:014058. DOI: 10.1103/PhysRevApplied.11.014058

[17] Lin TT, Gao Q, Liu GD, Dai XF, Zhang XM, Zhang HB. Dynamical stability, electronic and thermoelectric properties of quaternary ZnFeTiSi Heusler compound. Current Applied Physics. 2019;**19**:721-727. DOI: 10.1016/j.cap.2019.03.020

[18] Enamullah Cha P-R. The n- and p-type thermoelectric response of a semiconducting Co-based quaternary Heusler alloy: A density functional approach. Journal of Materials Chemistry C. 2019;**7**:7664-7671. DOI: 10.1039/c9tc00570f

[19] Ouardi S, Fecher GH, Felser C, Kübler J. Realization of spin gapless semiconductors: The Heusler compound Mn_2CoAl. Physical Review Letters. 2013;**110**:100401. DOI: 10.1103/PhysRevLett.110.100401

[20] Yousuf S, Gupta DC. Thermoelectric and mechanical properties of gapless Zr_2MnAl compound. Indian Journal of Physics. 2017;**91**:33-41. DOI: 10.1007/s12648-016-0900-3

[21] Patel PD, Shinde S, Gupta SD, Dabhi SD, Jha PK. The first principle calculation of structural, electronic, magnetic, elastic, thermal and lattice dynamical properties of fully compensated ferrimagnetic spin-gapless heusler alloy Zr_2MnGa. Computational Condensed Matter. 2018;**15**:61-68. DOI: 10.1016/j.cocom.2018.02.003

[22] Chen X, Huang Y, Yuan H, Liu J, Chen H. Theoretical investigation on thermoelectric properties of spin gapless semiconductor Cr_2ZnSi. Applied Physics A: Materials Science & Processing. 2018; **124**:841. DOI: 10.1007/s00339-018-2259-0

[23] Rani D, Enamullah, Bainsla L, Suresh KG, Alam A. Spin-gapless semiconducting nature of Co-rich $Co_{1+x}Fe_{1-x}CrGa$. Physical Review B. 2019;**99**:104429. DOI: 10.1103/PhysRevB.99.104429

[24] Yousuf S, Hien ND, Batoo KM, Raslan EH, Gupta DC. Insight into structural, electronic and thermoelectric properties of Zr_2MnX (X=Ga, In) Heuslers. Materials Research Express. 2019;**6**:046530. DOI: 10.1088/2053-1591/aafb1d

[25] Patel PD, Shinde SM, Gupta SD, Jha PK. A promising thermoelectric response of fully compensated ferrimagnetic spin gapless semiconducting Heusler alloy Zr_2MnAl at high temperature: DFT study. Materials Research Express. 2019;**6**: 076307. DOI: 10.1088/2053-1591/ab1723

[26] Patel PD, Shinde S, Gupta SD, Jha PK. Investigation of structural and elastic stability, electronic, magnetic, thermoelectric, lattice-dynamical and thermodynamical properties of spin gapless semiconducting Heusler alloy Zr_2MnIn using DFT approach. Journal of Electronic Materials. 2019;**48**: 1634-1642. DOI: 10.1007/s11664-018-06911-y

[27] Kuo CN, Lee HW, Wei C-M, Lin YH, Kuo YK, Lue CS. Ru_2NbGa: A Heusler-type compound with semimetallic characteristics. Physical Review B. 2016;**94**:205116. DOI: 10.1103/PhysRevB.94.205116

[28] Mondal S, Mazumdar C, Ranganathan R, Alleno E, Sreeparvathy PC, Kanchana V, et al. Ferromagnetically correlated clusters in semimetallic Ru_2NbAl Heusler alloy and its thermoelectric properties. Physical Review B. 2018;**98**:205130. DOI: 10.1103/PhysRevB.98.205130

[29] Ramachandran B, Lin YH, Kuo YK, Kuo CN, Gippius AA, Lue CS. Thermoelectric properties of Heusler-type $Ru_2VAl_{1-x}Ga_x$ alloys.

Intermetallics. 2018;**92**:36-41. DOI: 10.1016/j.intermet.2017.09.012

[30] Fichtner T, Kreiner G, Chadov S, Fecher GH, Schnelle W, Hoser A, et al. Magnetic and transport properties in the Heusler series $Ni_{2-x}Mn_{1+x}Sn$ affected by chemical disorder. Intermetallics. 2015; **57**:101-112. DOI: 10.1016/j.intermet.2014.10.012

[31] Khandy SA, Islam I, Gupta DC, Laref A. Predicting the electronic structure, magnetism, and transport properties of new Co-based Heusler alloys. International Journal of Energy Research. 2018;**42**:4221-4228. DOI: 10.1002/er.4182

[32] Khelfaoui F, Boudali A, Bentayeb A, El Hachemi OL, Si AY. Investigations of structural, elastic, electronic, magnetic and transport properties of the heusler compounds Zr_2PdZ (Z = Al, Ga, and In): FP-LAPW method. Acta Physica Polonica A. 2018;**133**:157-163. DOI: 10.12693/APhysPolA.133.157

[33] Benzoudji F, Bensaid D, el Kader YA, Arbouche O, Bouyakoub AZ, Moulay N, et al. The preference of the ferromagnetic ordering for the novel Heusler Rh_2MnTi compound. Journal of Superconductivity and Novel Magnetism. 2019;**32**: 1415-1421. DOI: 10.1007/s10948-018-4837-y

[34] Duan J, Yin-Wei W, A-Peng Z, Liu S, Dar SA. Electronic structure, elastic, mechanical, thermodynamic and thermoelectric investigations of Mn_2PtX (X=Rh, Pd) Heusler alloys. Solid State Communications. 2019;**290**:12-21. DOI: 10.1016/j.ssc.2018.12.013

[35] Galanakis I, Özdoğan K, Şaşıoğlu E, Aktaş B. Ab initio design of half-metallic fully compensated ferrimagnets: The case of Cr_2MnZ (Z=P, As, Sb, and Bi). Physical Review B. 2007;**75**:172405. DOI: 10.1103/PhysRevB.75.172405

[36] Graf T, Fecher GH, Barth J, Winterlik J, Felser C. Electronic structure and transport properties of the Heusler compound Co_2TiAl. Journal of Physics D: Applied Physics. 2009;**42**: 084003. DOI: 10.1088/0022-3727/42/8/084003

[37] Balke B, Ouardi S, Graf T, Barth J, Blum CGF, Fecher GH, et al. Seebeck coefficients of half-metallic ferromagnets. Solid State Communications. 2010;**150**:529-532. DOI: 10.1016/j.ssc.2009.10.044

[38] Barth J, Fecher GH, Balke B, Ouardi S, Graf T, Felser C, et al. Itinerant half-metallic ferromagnets Co_2TiZ (Z=Si, Ge, Sn): Ab initio calculations and measurement of the electronic structure and transport properties. Physical Review B. 2010;**81**: 064404. DOI: 10.1103/PhysRevB.81.064404

[39] Sharma V, Solanki AK, Kashyap A. Electronic, magnetic and transport properties of Co_2TiZ (Z=Si, Ge and Sn): A first-principle study. Journal of Magnetism and Magnetic Materials. 2010;**322**:2922-2928. DOI: 10.1016/j.jmmm.2010.05.006

[40] Graf T, Barth J, Balke B, Populoh S, Weidenkaff A, Felser C. Tuning the carrier concentration for thermoelectrical application in the quaternary Heusler compound $Co_2TiAl_{1-x}Si_x$. Scripta Materialia. 2010; **63**:925-928. DOI: 10.1016/j.scriptamat.2010.07.001

[41] Graf T, Barth J, Blum CGF, Balke B, Felser C. Phase-separation-induced changes in the magnetic and transport properties of the quaternary Heusler alloy $Co_2Mn_{1-x}Ti_xSn$. Physical Review B. 2010;**82**:104420. DOI: 10.1103/PhysRevB.82.104420

[42] Barth J, Fecher GH, Balke B, Graf T, Shkabko A, Weidenkaff A, et al. Anomalous transport properties of the

half-metallic ferromagnets Co_2TiSi, Co_2TiGe and Co_2TiSn. Philosophical Transactions of the Royal Society A. 2011;**369**:3588-3601. DOI: 10.1098/rsta.2011.0183

[43] Schwall M, Schoop LM, Ouardi S, Balke B, Felser C, Klaer P, et al. Thermomagnetic properties improved by self-organized flower-like phase separation of ferromagnetic $Co_2Dy_{0.5}Mn_{0.5}Sn$. Advanced Functional Materials. 2012;**22**:1822-1826. DOI: 10.1002/adfm.201102792

[44] Sharma S, Pandey SK. Investigation of thermoelectric properties of half-metallic Co_2MnGe by using first principles calculations. Journal of Physics: Condensed Matter. 2014;**26**: 215501. DOI: 10.1088/0953-8984/26/21/215501

[45] Mohankumar R, Ramasubramanian S, Rajagopalan M, Manivel Raja M, Kamat SV, Kumar J. Effect of Fe substitution on the electronic structure, magnetic and thermoelectric properties of Co_2FeSi full Heusler alloy: A first principle study. Computational Materials Science. 2015; **109**:34-40. DOI: 10.1016/j.commatsci.2015.07.001

[46] Sharma S, Pandey SK. Applicability of two-current model in understanding the electronic transport behavior of inverse Heusler alloy: Fe_2CoSi. Physics Letters A. 2015;**379**:2357-2361. DOI: 10.1016/j.physleta.2015.04.019

[47] Bhat IH, Yousuf S, Mohiuddin Bhat T, Gupta DC. Investigation of electronic structure, magnetic and transport properties of half-metallic Mn_2CuSi and Mn_2ZnSi Heusler alloys. Journal of Magnetism and Magnetic Materials. 2015;**395**:81-88. DOI: 10.1016/j.jmmm.2015.07.022

[48] Bhat TM, Gupta DC. Transport, structural and mechanical properties of quaternary FeVTiAl alloy. Journal of Electronic Materials. 2016;**45**: 6012-6018. DOI: 10.1007/s11664-016-4827-4

[49] Yalcin BG. Ground state properties and thermoelectric behavior of Ru_2VZ (Z=Si, Ge, Sn) half-metallic ferromagnetic full-Heusler compounds. Journal of Magnetism and Magnetic Materials. 2016;**408**:137-146. DOI: 10.1016/j.jmmm.2016.02.064

[50] Reshak AH. Transport properties of Co-based Heusler compounds Co_2VAl and Co_2VGa: Spin-polarized DFT+U. RSC Advances. 2016;**6**:54001-54012. DOI: 10.1039/c6ra10226c

[51] Bhat TM, Gupta DC. Robust thermoelectric performance and high spin polarisation in CoMnTiAl and FeMnTiAl compounds. RSC Advances. 2016;**6**:80302. DOI: 10.1039/C6RA18934B

[52] Popescu V, Kratzer P, Wimmer S, Ebert H. Native defects in the Co_2TiZ (Z=Si,Ge,Sn) full Heusler alloys: Formation and influence on the thermoelectric properties. Physical Review B. 2017;**96**:054443. DOI: 10.1103/PhysRevB.96.054443

[53] Hayashi K, Eguchi M, Miyazaki Y. Structural and thermoelectric properties of ternary full-Heusler alloys. Journal of Electronic Materials. 2017;**46**:2710-2716. DOI: 10.1007/s11664-016-4944-0

[54] Yousuf S, Gupta DC. Insight into mechanical properties and thermoelectric efficiency of Zr_2CoZ (Z = Si, Ge) Heusler alloys. Materials Research Express. 2017;**4**(11):6307. DOI: 10.1088/2053-1591/aa96cb

[55] Bhat TM, Gupta DC. Effect of on-site coulomb interaction on electronic and transport properties of 100% spin polarized CoMnVAs. Journal of Magnetism and Magnetic Materials. 2017;**435**:173-178. DOI: 10.1016/j.jmmm.2017.04.012

[56] Yousuf S, Khandy SA, Bhat TM, Gupta DC. Evaluation of mechanical and transport properties of Zr_2CoSi Heusler alloy. AIP Conf. Proc. 2017; **1832**:110005. DOI: 10.1063/1.4980629

[57] Yousuf S, Gupta DC. Investigation of electronic, magnetic and thermoelectric properties of Zr_2NiZ (Z = Al,Ga) ferromagnets. Materials Chemistry and Physics. 2017;**192**:33-40. DOI: 10.1016/j.matchemphys.2017.01.056

[58] Patel PD, Pillai SB, Shinde SM, Gupta SD, Jha PK. Electronic, magnetic, thermoelectric and lattice dynamical properties of full heusler alloy Mn_2RhSi: DFT study. Physica B: Condensed Matter. 2018;**550**:376-382. DOI: 10.1016/j.physb.2018.09.020

[59] Gupta DC, Ghosh S. High-pressure and temperature dependence of electronic, magnetic, elastic, thermodynamic, and transport properties of full-Heusler alloys Co_2YIn (Y = Nb, Zr). Journal of Superconductivity and Novel Magnetism. 2018;**31**:2465-2483. DOI: 10.1007/s10948-017-4498-2

[60] Li H, Hayashi K, Miyazaki Y. Design and fabrication of full-Heusler compound with positive Seebeck coefficient as a potential thermoelectric material. Scripta Materialia. 2018;**150**: 130-133. DOI: 10.1016/j.scriptamat.2018.03.018

[61] Yousuf S, Gupta DC. Chemical potential evaluation of thermoelectric and mechanical properties of Zr_2CoZ (Z = Si, Ge) Heusler alloys. Journal of Electronic Materials. 2018;**47**: 2468-2478. DOI: 10.1007/s11664-017-6034-3

[62] Hossain MA, Rahman MT, Khatun M, Haque E. Structural, elastic, electronic, magnetic and thermoelectric properties of new quaternary Heusler compounds CoZrMnX (X=Al, Ga, Ge, In). Computational Condensed Matter. 2018;**15**:31-41. DOI: 10.1016/j.cocom.2018.03.006

[63] Enamullah, Lee S-C. Robust mechanical stability, electronic structure, magnetism and thermoelectric properties of CoFeMnSb quaternary Heusler alloy: A first principle study. Journal of Alloys and Compounds. 2018;**742**:903-909. DOI: 10.1016/j.jallcom.2018.01.330

[64] Bhat TM, Gupta DC. Magneto-electronic, thermal, and thermoelectric properties of some Co-based quaternary alloys. Journal of Physics and Chemistry of Solids. 2018;**112**:190-199. DOI: 10.1016/j.jpcs.2017.09.023

[65] Sunmonu RS, Akinlami JO, Dare EO, Adebayo GA. Effects of Y atom substitution on the structural, magnetic, electronic, elastic, mechanical, thermodynamic and thermoelectric properties of Co_2YAl (Y = Cr, Mn) full Heusler alloys from first principles investigations. Computational Condensed Matter. 2019;**21**:e00412. DOI: 10.1016/j.cocom.2019.e00412

[66] Agbaoye RO, Adebambo PO, Adetunji BI, Osafile O, Adebayo GA. Thermoelectric properties, optimal doping levels and high figure of merit in cobalt-based half/full Heusler alloys by first-principles calculations. Materials Science and Engineering B. 2019;**248**:114409. DOI: 10.1016/j.mseb.2019.114409

[67] Koshi NA, John R. First principles study on the structural, electronic, magnetic and thermoelectric properties of CoX'NbGa (X' = Cr, Mn, Fe) quaternary Heusler alloys. European Physical Journal B. 2019;**92**:86. DOI: 10.1140/epjb/e2019-90663-3

[68] Boudali A, Mokaddem A, Doumi B, Moujri H. Using the spin–orbit coupling

for studying the structural, elastic, electronic, ferromagnetic, and thermoelectric properties of the Ti_2PdX (X = Al, Ga, and in) full-Heusler alloy compounds. Acta Physica Polonica A. 2019;**135**:409-419. DOI: 10.12693/APhysPolA.135.409

[69] Alrahamneh MJ, Mousa AA, Khalifeh JM. First principles study of the structural, electronic, magnetic and thermoelectric properties of Zr_2RhAl. Physica B: Condensed Matter. 2019;**552**: 227-235. DOI: 10.1016/j.physb.2018.10.018

[70] Singh S, Gupta DC. Electronic structure, mechanical, thermoelectric, optical, and thermodynamic properties of yttrium-based quaternary Heusler alloys. International Journal of Energy Research. 2019;**43**:8633-8648. DOI: 10.1002/er.4860

[71] Singh S, Gupta DC. Lanthanum based quaternary Heusler alloys LaCoCrX (X = Al, Ga): Hunt for half-metallicity and high thermoelectric efficiency. Results in Physics. 2019;**13**: 102300. DOI: 10.1016/j.rinp.2019.102300

[72] Chatterjee S, Das S, Pramanick S, Chatterjee S, Giri S, Banerjee A, et al. Anomalous transport and magnetic behaviours of the quaternary Heusler compounds CoFeTiSn and CoFeVGa. Journal of Magnetism and Magnetic Materials. 2019;**478**:155-160. DOI: 10.1016/j.jmmm.2019.01.100

[73] Khandy SA, Chai J-D. Robust stability, half-metallic ferrimagnetism and thermoelectric properties of new quaternary Heusler material: A first principles approach. Journal of Magnetism and Magnetic Materials. 2020;**502**:166562. DOI: 10.1016/j.jmmm.2020.166562

[74] Hoat DM, Naseri M. Examining the half-metallicity and thermoelectric properties of new equiatomic quaternary Heusler compound CoVRhGe under pressure. Physica B: Condensed Matter. 2020; **583**:412058. DOI: 10.1016/j.physb.2020.412058

[75] Sofi SA, Gupta DC. High pressure-temperature study on thermodynamics, half-metallicity, transport, elastic and structural properties of Co-based Heusler alloys: A first-principles study. Journal of Solid State Chemistry. 2020; **284**:121178. DOI: 10.1016/j.jssc.2020.121178

[76] Sofi SA, Gupta DC. Exploration of electronic structure, mechanical stability, magnetism, and thermophysical properties of L21 structured Co_2XSb (X = Sc and Ti) ferromagnets. International Journal of Energy Research. 2020;**44**:2137-2149. DOI: 10.1002/er.5071

[77] Ilkhani M, Boochani A, Amiri M, Asshabi M, Rai DP. Mechanical stability and thermoelectric properties of the PdZrTiAl quaternary Heusler: A DFT study. Solid State Communications. 2020;**308**:113838. DOI: 10.1016/j.ssc.2020.113838

[78] Li H, Hayashi K, Dong J, Li J-F, Miyazaki Y. Distinct impact of order degree on thermoelectric power factor of p-type full-Heusler Mn_2VAl compounds. Materials Research Express. 2020;**7**:055503. DOI: 10.1088/2053-1591/ab875b

[79] Galanakis I, Dederichs PH, Papanikolaou N. Slater-Pauling behavior and origin of the half-metallicity of the full-Heusler alloys. Physical Review B. 2002;**66**:174429. DOI: 10.1103/PhysRevB.66.174429

[80] Skaftouros S, Özdoğan K, Şaşıoğlu E, Galanakis I. Generalized Slater-Pauling rule for the inverse Heusler compounds. Physical Review B.

2013;**87**:024420. DOI: 10.1103/PhysRevB.87.024420

[81] Özdoğan K, Şaşıoğlu E, Galanakis I. Slater-Pauling behavior in LiMgPdSn-type multifunctional quaternary Heusler materials: Half-metallicity, spin-gapless and magnetic semiconductors. Journal of Applied Physics. 2013;**113**:193903. DOI: 10.1063/1.4805063

[82] Allen PB. Boltzmann theory and resistivity of metals. In: Chelikowsky JR, Louie SG, editors. Quantum Theory of Real Materials. Boston: Kluwer; 1996. pp. 219-250

[83] Ziman JM. Electrons and phonons. In: Oxford Classics Series. Oxford: Clarendon Press; 2001. p. 568

[84] Madsen GKH, Singh DJ. BoltzTraP. A code for calculating band-structure dependent quantities. Computer Physics Communications. 2006;**175**:67-71. DOI: 10.1016/j.cpc.2006.03.007

[85] Mott NF, Davis EA. Electronic Processes in Non-Crystalline Materials. Oxford Classic Texts in the Physical Sciences. Oxford: Oxford University Press; 2012. p. 608

[86] Zhang Y, Jia X, Sun H, Sun B, Liu B, Liu H, et al. Suppressing adverse intrinsic conduction of Bi_2Te_3 thermoelectric bulks by Sb and Cu co-substitutions via HPHT synthesis. RSC Advances. 2016;**6**:7378-7383. DOI: 10.1039/C5RA24236C

[87] May AF, Snyder GJ. Introduction to modeling thermoelectric transport at high temperatures. In: Rowe DM, editor. Thermoelectrics and its Energy Harvesting. London: CRC Press; 2012. p. 1

[88] Tong Z, Li S, Ruan X, Bao H. Comprehensive first-principles analysis of phonon thermal conductivity and electron-phonon coupling in different metals. Physical Review B. 2019;**100**:144306. DOI: 10.1103/PhysRevB.100.144306

[89] Tsidilkovski IM. Electron Spectrum of Gapless Semiconductors. Springer Series in Solid-State Sciences. Vol. 116. New York: Springer Berlin; 1997. p. 249. DOI: 10.1007/978-3-642-60403-4

[90] Galanakis I, Özdoğan K, Şaşıoğlu E, Blügel S. Conditions for spin-gapless semiconducting behavior in Mn_2CoAl inverse Heusler compound. Journal of Applied Physics. 2014;**115**:093908. DOI: 10.1063/1.4867917

[91] Xu XD, Chen ZX, Sakuraba Y, Perumal A, Masuda K, Kumara LSR, et al. Microstructure, magnetic and transport properties of a Mn_2CoAl Heusler compound. Acta Materialia. 2019;**176**:33-42. DOI: 10.1016/j.actamat.2019.06.047

[92] Cahill DG, Pohl RO. Lattice vibrations and heat transport in crystals and glasses. Annual Review of Physical Chemistry. 1988;**39**:93-121. DOI: 10.1146/annurev.pc.39.100188.000521

[93] Bharwdaj A, Jat KS, Patnaik S, Parkhomenko YN, Nishino Y, Khovaylo VV. Current research and future prospective of iron-based Heusler alloys as thermoelectric materials. Nanotechnologies in Russia. 2019;**14**:281-289. DOI: 10.1134/S1995078019040049

[94] Hinterleitner B, Knapp I, Poneder M, Shi Y, Müller H, Eguchi G, et al. Thermoelectric performance of a metastable thin-film Heusler alloy. Nature. 2019;**576**:85-90. DOI: 10.1038/s41586-019-1751-9

[95] Oliynyk AO, Antono E, Sparks TD, Ghadbeigi L, Gaultois MW, Meredig B, et al. High-throughput machine-learning-driven synthesis of full-Heusler compounds. Chemistry of Materials. 2016;**28**:7324-7331. DOI: 10.1021/acs.chemmater.6b02724

[96] Kim K, Ward L, He J, Krishna A, Agrawal A, Wolverton C. Machine-learning-accelerated high-throughput materials screening: Discovery of novel quaternary Heusler compounds. Physical Review Materials. 2018;**2**: 123801. DOI: 10.1103/PhysRevMaterials.2.123801

Chapter 5

Zr-Based Heusler Compounds for Biomedical Spintronic Applications

Anca Birsan and Victor Kuncser

Abstract

Current advances in microelectronics depend on novel approaches based on the synergistic use of charge and spin dynamics of electrons in multi-functional materials. Such new concepts have already found practical applications in magnetoelectronics or spintronics (e.g., spin valves or nonvolatile memory components). For efficient spintronic devices, it is desirable to have an enhanced spin polarization, that, to work with nearly 100% spin-polarized currents. Since half metallic materials have electrons of unique spin polarization around the Fermi level (finite density of states in only one spin channel), they are promising candidates for use as spin injectors in spintronic devices. Although the Heusler compounds reported in the literature presenting half-metallic ferro/ferrimagnetism are numerous, only a few contain elements with low toxicity, as for example zirconium, being also susceptible of convenient preparation and processing. Therefore, in future, zirconium-based compounds could become a much suitable alternative to the presently known cobalt, iron, chromium, titanium, manganese, or scandium-based half-metallic Heusler compounds, being of interest especially in biomedical spintronic related applications involved in corrosive/active environment.

Keywords: Heusler compounds, spintronics, half metallic, spin gapless semiconductors

1. Introduction

The recent developments in thin films and nanofabrication techniques of biosensors and related spintronic devices are the forefront of current research efforts, bridging material sciences, physics, chemistry, and engineering, to form a seamless integration of digital world into the soft or living systems. Magnetic functionalities may provide a sense of proximity, orientation, or displacement to this novel formulation of biomedical electronics.

High spin polarization is one of the requested and necessary properties of materials used as electrodes or spin pumping/spin analyzers elements in spintronics, including those used in medicine and this is by definition a characteristic of alloys with half-metallic properties [1, 2]. The property of half-metallic ferromagnetism initially discovered in Cu_2MnAl compound [2] consists in a metallic behavior of one spin channel of electronic structure and a semiconducting one in the other, thus creating a material with hybrid properties between metals and semiconductors. As

 IntechOpen

a direct consequence, there will be always a 100% net spin polarization at the Fermi level due to the unique spin polarization of electrons in only one channel.

In materials in which the unit cell consists of two distinct sublattices with antiferromagnetic coupling between them, an internal spin partial compensation occurs and this particular property was referred as half-metallic ferrimagnetism [3, 4], which comparing to half-metallic ferromagnetism exhibits lower magnetic moments per formula unit (f.u) and weaker stray fields. Moreover, if the magnetic moments of the constituent sublattices fully compensate each other (with a net spin = 0 μ_B/f.u.), an alloy with a completely compensated ferrimagnetism (CCF) [5] resulted and the compound was classified as half-metallic completely compensated ferrimagnet (HM-CCFs) [6]. However, such a complete spin polarization of carriers occurs in the case of zero temperature and only in the absence of the spin-orbit interactions. Apart from this, HM-CCFs are intensively studied to develop new stable spin-polarized electrodes for biomedical in-vivo applications, junctions or integrated spin-transfer torque nano-oscillators for telecommunication.

A particular class of half-metallic materials is Spin Gapless Semiconductors (SGS). These compounds exhibit around Fermi level, in one spin channel a typical semiconducting band gap, while in the other (where in usual half-metallic compounds a metallic character is present), the negligible density of states are equivalent to a very narrow almost zero band gap. The above described characteristic of electronic structure, places SGSs at the boundary between half-metallic compounds and semiconductors.

In this particular case of half-metallicity, the materials act like topological insulators, where in particular high Curie temperature may coexist with high resistance. A combination of spin gapless semiconducting properties with completely compensated ferrimagnetism (0 μ_B total magnetic moment per f.u.) leads to spin gapless completely compensated ferrimagnetism (SG-CCF) [7].

Particularly, attractive classes of alloys exhibiting half-metallic properties, based on which may be developed biosensors, the new electrode materials with high spin polarization include alloys like Heusler compounds [2]. This class of materials, used in present as electrodes for magnetic tunnel junctions were discovered by Fritz Heusler, in 1903, who reported that Cu_2MnAl alloy is ferromagnetic, even though, alone, none of constituent elements has magnetic properties [8]. These intermetallic alloys are described by two variants: the half-Heusler XYZ compounds, with C_{1b} crystal structure and the full-Heusler X_2YZ variants which typically crystallize in Cu_2MnAl (cubic L_{21})-type structure; where X is a transition metal, Y may be a rare-metal or a transition metal, and Z is a main group element. Recently, it has been shown that in case of a full-Heusler compounds X_2YZ, if the Y element is more electronegative than X, a structure with Hg_2CuTi-prototype is observed. This is the so called as inverse Heusler structure, crystallizing in F43m space group [9], with X atoms placed in the 4a(0,0,0) and 4c(1/4,1/4,1/4) Wyckoff positions, Y in 4b(1/2,1/2,1/2) and Z in 4d(3/4,3/4,3/4), respectively. In this crystal structure, no octahedral symmetry O_h is adopted, and all atoms have tetrahedral symmetry T_d.

The Slater-Pauling curve gives the interrelation between the total magnetic moment and the valence electron concentration in ferromagnetic/ferrimagnetic alloys [10, 11]. The original Slater-Pauling approach suggests the existence of different laws, due to the average over all atoms of the total magnetic moment and the number of valence electrons. For compounds with different kinds of atoms and ordered crystalline structures, it is more appropriate to consider all atoms of the unit cell, to find the magnetic moment per unit cell.

In terms of two-orbital two-electron stabilizing interactions, within the framework of density functional theory, the states of each spin channel are occupied according to several aspects concerning ionic arguments, crystal structure of primitive cell, lattice parameter, approximations made for the exchange and correlation interaction, energy threshold set between the core and valence states, and also Brillouin zone integration mesh. Based on ionic arguments, the most electropositive element transfers the valence electrons to the most electronegative element. The purpose is to obtain stable closed shell ions. In addition, strongly dependent by the atomic arrangement of atoms and environment, hybridization occurs whenever the sum of metallic radii (12-coordinated) of two first-neighbors exceeds the interatomic distance.

A particularly useful measure to describe the electronic properties of a material is the electron spin polarization P at Fermi level (ε_F), given by Eq. (1)

$$P = \frac{\rho_\uparrow(\varepsilon_F) - \rho_\downarrow(\varepsilon_F)}{\rho_\uparrow(\varepsilon_F) + \rho_\downarrow(\varepsilon_F)} \tag{1}$$

where $\rho_\uparrow(\varepsilon_F)$ and $\rho_\downarrow(\varepsilon_F)$ denote the spin projected density of states around Fermi energy. The states of opposite spin (majority and minority spin states or spin-up and spin-down states) are represented by arrows $\uparrow$ and $\downarrow$. Depending on the magnetic characteristic of the material, the electron spin polarization vanishes in case of antiferromagnetic and paramagnetic compounds or has a finite value for ferrimagnetic and ferromagnetic alloys, below the Curie temperature. When either $\rho_\uparrow(\varepsilon_F)$ or $\rho_\downarrow(\varepsilon_F)$ equals zero, the electrons around Fermi level are fully spin polarized.

For ternary 1:1:1 Heusler compounds, the Slater-Pauling rule was firstly reported by Kübler [12]. These compounds, with C_{1b} structure have three atoms per unit cell and follow the Slater-Pauling 18-electron-rule ($M_t = Z_t - 18$), where M_t is the total magnetic moment per the formula unit, Z_t is the total number of valence electrons, and 18 represents the number of occupied states in the spin bands. A Slater-Pauling 24-electron-rule ($M_t = Z_t - 24$) was found for the 2:1:1 family of full-Heusler compounds with L_{21} structure (Cu_2MnAl-prototype) [13]. The present work deals only with ternary 2:1:1 full-Heusler compounds with Hg_2CuTi type structure. Even though the origin of the band gap in the latter 2:1:1 full-Heusler compounds is different than that of the ternary 1:1:1 Heusler compounds, the corresponding Slater-Pauling rule is similar: 18-electron-rule ($M_t = Z_t - 18$). This Slater-Pauling 18-electron-rule was recently explained for Ti_2-based full-Heusler compounds [4, 14].

Many Co_2, Mn_2, Ti_2, and Sc_2 – Heusler compounds reported in literature are ferromagnetic [15–18], ferrimagnetic half-metals [19], or spin gapless semiconductors [20]. Among them, Mn_2CoAl full-Heusler compound crystallizing in Hg_2CuTi-prototype was extensively studied: theoretically investigated, the structure was experimentally verified by XRD and the electron transport characteristics where obtained by a Physical Properties Measurement System (PPMS) on samples cut from ingots. The total magnetic moment was experimentally measured using a Magnetic Properties Measurement System (MPMS) [20]. Zirconium has a Pauling electronegativity value lower than those of all d-elements and hence Zr-based Heusler materials are supposed to crystallize in Hg_2CuTi type structure, similar to Mn_2CoAl.

Zirconium-based Heusler compounds were selected because they exhibit low toxicity and are corrosion resistant, being therefore susceptible of convenient preparation and processing in the field of electronic biomedical sensors ranging from healthcare and medical diagnosis, food safety, and environmental monitoring to life science research.

The information about the experimental preparation and electronic structure of Zr-based Heusler compounds with true half-metallic properties are still scarce. Therefore, to understand the properties of potential zirconium-based Heusler compounds, in the beginning, theoretical investigations can be performed via density functional theory (DFT). Self-consistent calculations using a "muffin-tin" model and various approximations to describe the exchange and correlation interactions can lead to valuable information about the energetically favorable crystalline structure, electronic configuration, or magnetic properties by means of the total energy minimization.

This chapter gives a comprehensive overview of the key electronic structures and magnetic properties usually found in half-metallic zirconium-based full-Heusler compounds.

2. Half-metallic ferrimagnetic materials

The cubic crystal structure of full-Heusler Zr_2YZ variants exhibits two magnetic sublattices, coupled to each other. Thus, the two Zr atoms are located in tetrahedral lattice sites and interact to each other. In addition, Zr and Y atoms form a second and more delocalized magnetic sublattice. Therefore, ferrimagnetic interaction between the Zr and Y atoms is frequently reported phenomena.

The total spin-polarized density of states of a typical half metallic ferrimagnetic material exhibits in the spin-up channel a semiconducting band gap while in the spin-down channel a metallic behavior. A relevant example is illustrated in **Figure 1** for Zr_2CrAl (unpublished results). The main contribution to the total density of states from spin-down channel comes from transition metal constituent elements, and these results are consistent with other published information [21].

Desirable candidates for magnetoelectronic devices, half-metallic ferrimagnetic compounds provide an unequivocal advantage over their ferromagnetic counterparts by reduction of the magnetic moment due to the ferrimagnetic interaction resulted from compensation of partial magnetic moments of the two different magnetic sublattices. This phenomenon is illustrated in **Figure 2** (unpublished results) for the Zr_2CrAl compound, where one can notice the magnetic moment of Cr atoms, partially compensated by magnetic moments of Zr located in the two different sublattices and having different neighborhoods. Similar DFT outcome were reported for Zr_2YZ (Y = Cr, V, Z = Al, Ga, In, Pb, Sn, Tl) [21–24].

Figure 3 (unpublished results) exhibits the position of the Fermi level and the width of the energy gap in spin-up channel as function of the lattice parameter. According to theoretical investigations, the Zr_2CrAl compound is a potential ideal candidate for spintronics, due to the presence of a steady energy gap in only one spin channel, for a large lattice parameter range.

Table 1 summarizes the published results regarding Zr_2CrZ (Z = Al, Ga, In) [21, 22]. As can be seen, the energy band gap (E_g) from spin-up channel increases as the atomic radii of Z elements increase. The ferrimagnetic interaction occurs between the zirconium atoms from both sublattice and the chromium ones, phenomena which are reflected by the opposite sign of the partial magnetic moments of Zr and Cr atoms. The total magnetic moment per f.u. for all compounds strictly follow the Slater Pauling rule described earlier.

In the ideal case of a fully compensated magnetic moment, a half-metallic ferromagnetic material would be obtained, useful to be applied in a junction device as a stable spin-polarized electrode based on spin-transfer effect.

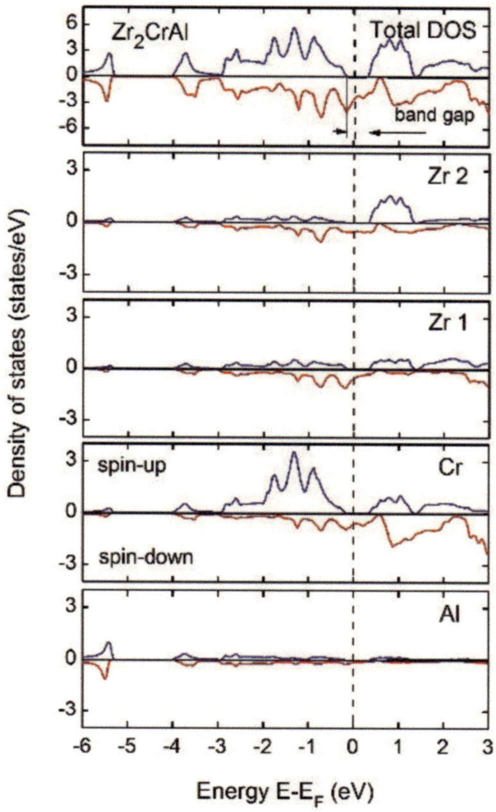

Figure 1.
Partial and total density of states (PDOS and TDOS) of half-metallic ferrimagnetic Heusler compound, Zr_2CrAl at optimized lattice parameter.

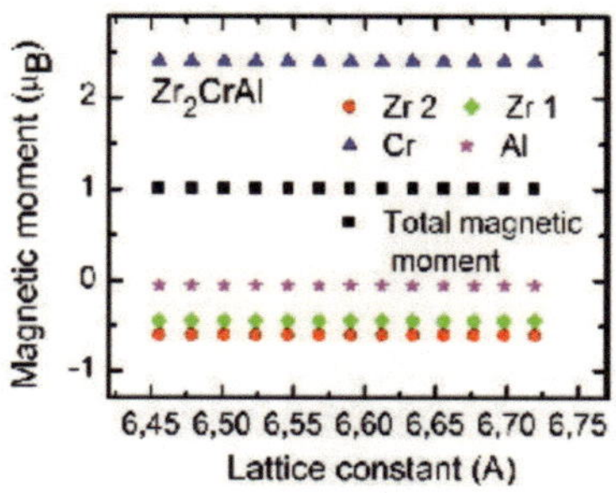

Figure 2.
Partial and total magnetic moments in Zr_2CrAl Heusler compound.

The theoretical results from in literature for Zr_2VZ (Z = Al, Ga, In, Si, Ge, Sn, Pb) [23, 24] report that the most energetically favorable crystalline structure comparing with the Hg_2CuTi structure has the prototype Cu_2MnAl and in this configuration the materials do not present half-metallic properties. However, the Hg_2CuTi type structure can be synthesized experimentally due to the negative entropy of formation. In the inverse Heusler crystalline structure, the Zr_2VZ exhibits half metallic ferrimagnetic characteristics, the partial magnetic moment of Vanadium being opposite to the one of zirconium atoms.

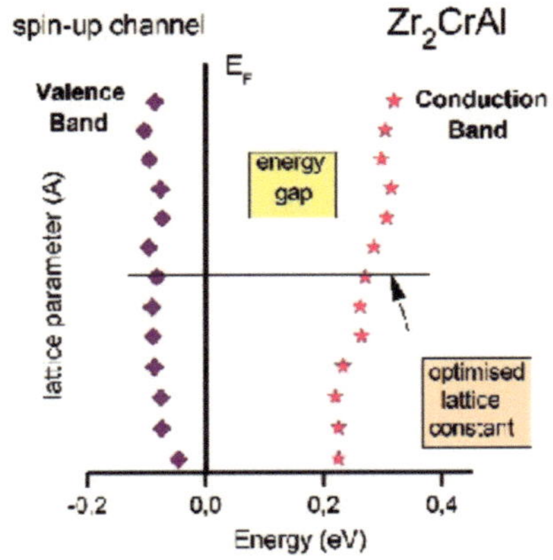

Figure 3.
The positions of the highest occupied states from the valence band (solid rhombs) and of the lowest unoccupied states from the conduction band (solid stars) of total DOSs (spin-up channel) for Zr_2CrAl as function of the lattice parameter.

Alloy	a (Å)	$\mu_{Zr(4a)}$ (μ_B/atom)	$\mu_{Zr(4c)}$ (μ_B/atom)	$\mu_{Y(4b)}$ (μ_B/atom)	$\mu_{Z(4d)}$ (μ_B/atom)	μ_t (μ_B/f.u.)	E_g (eV)
Zr_2CrAl	6.59[b]	-0.955[b]	-0.768[b]	2.835[b]	-0.110[b]	1.000[b]	0.452[b]
Zr_2CrGa	6.635[a]	0.849[a]	0.702[a]	-2.591[a]	0.049[a]	-1.000[a]	0.629[a]
	6.622[b]	-1.011[b]	-0.914[b]	2.994[b]	-0.068[b]	1.000[b]	0.512[b]
Zr_2CrIn	6.875[a]	1.016[a]	0.859[a]	-2.930[a]	0.034[a]	-1.000[a]	0.673[a]
	6.812[b]	-1.213[b]	-1.080[b]	3.343[b]	-0.049[b]	1.000[b]	0.615[b]

[a]*Ref. [22].*
[b]*Ref. [21].*

Table 1.
Calculated lattice parameters, partial, total magnetic moments, and energy band gap in Zr_2CrZ (Z = Al, Ga, In).

3. Spin gapless semiconductors

Particular cases of half-metallic ferromagnetic materials are the spin gapless semiconductors, where a semiconducting band gap is formed in one spin channel and a pseudo-band gap in the other one. Such a pseudo band gap is often called zero or closed band gap because the maximum energy of the valence band is very close to the minimum energy of the conduction band. The Zr_2MnAl compound presents a typical behavior of spin gapless semiconductors and may allow a tunable spin transport (see **Figure 4**). The Fermi level, located at 0.04 eV below the conduction band minimum, in case of Zr_2MnAl, falls into a typical spin gapless semiconducting band gap of 0.41 eV in spin-up channel, according to Ref [25]. In the spin-down channel, a zero band gap is reported around the Fermi level. In both spin channels, the significant contribution to density of states between -4.5 and -1.5 eV comes from the 3d electrons of Mn, while the 4d electrons from Zr atoms have contribution only above the Fermi level.

Figure 5 presents the contribution of double and triple degenerated states (d_{eg} and d_{t2g}, respectively) of Zr and Mn atoms, calculated around the Fermi level, at optimized lattice parameters. In Zr_2MnAl compound, the highest bonding states from valence band, below the E_F, belong to triple degenerated states of manganese

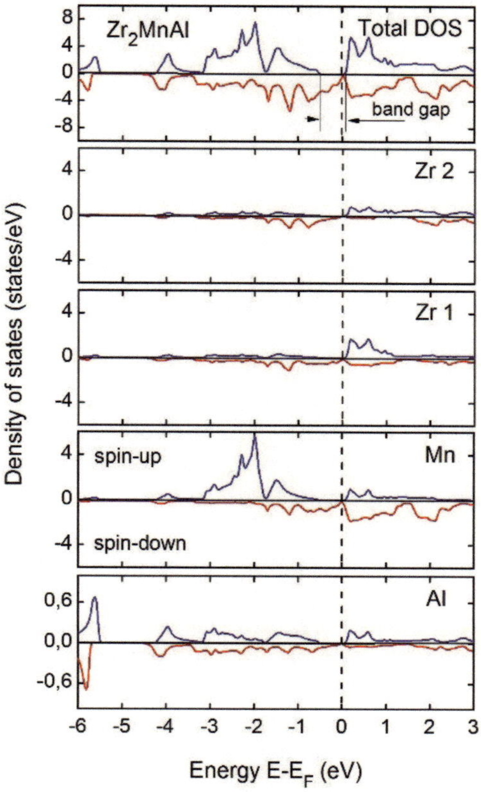

Figure 4.
Partial and total density of states (PDOS and TDOS) of spin gapless semiconductor Zr_2MnAl at equilibrium lattice parameter.

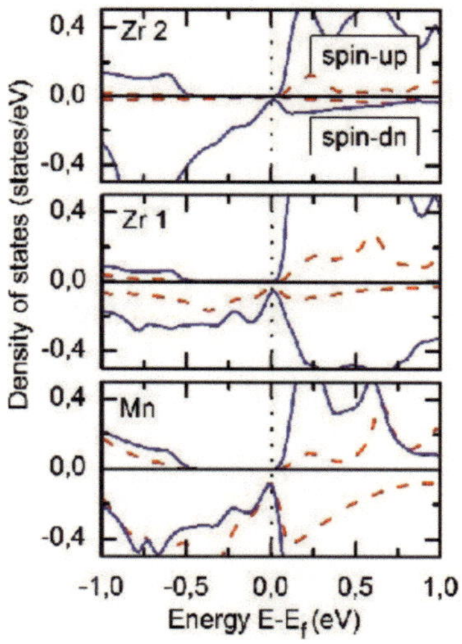

Figure 5.
The densities of states of double and triple degenerated states of Zr and Mn atoms, around the Fermi level, calculated at optimized lattice parameters for Zr_2MnAl. The Fermi level, d_{eg} and d_{t2g}, are illustrated with black dotted, red dashed, and blue solid line, respectively.

d_{t2g}, while the lowest anti-bonding states from conduction band come from the triple degenerated states d_{t2g} of Zr_1, Zr_2, and Mn. As a result, the energy gap from spin-up channel results due to Zr-Mn hybridization. The Zr_2MnAl alloy presents an indirect band gap of 0.41 eV, in the spin up channel with the higher bonding states from valence band located in the Δ point and the lowest anti-bonding states from the conduction band, distributed in the Δ and W high symmetry points of Brillouin zone.

It is obvious that the change in the lattice parameter affects the presence of the zero band gap from spin channel and the width of the semiconducting band gap. In Zr_2MnAl alloy, the band gap increases initially by increasing the lattice parameter. The largest band gap is obtained for a lattice parameter of 6.6 Å, which corresponds to a volume increase of 2%. Above the lattice parameter of 6.6 Å, the spin gapless semiconducting properties of Zr_2MnAl compound disappear, due to the shifts of the Fermi level in to the conduction band. The width of the energy band gap from spin-up channel decreases as illustrated in **Figure 6**.

The spin gapless semiconductors may present a finite total magnetic moment; however in the particular case, when a perfectly compensated ferrimagnetism appears the total magnetic moment of compound equals zero and the alloy becomes a spin gapless completely compensated ferrimagnet, like Zr_2MnAl (see **Figure 7**).

Surprisingly, the zirconium element which does not exhibit natively magnetic properties shows magnetic behavior. A ferrimagnetic interaction occurs between the magnetic moments of Zr and Mn atoms, whereas the zirconium atoms, located in different Wyckoff positions, are coupled ferromagnetically. The magnetic moments of manganese increase with the lattice parameter, in all compounds. The magnetic moments of zirconium atoms coupled ferromagnetically decrease with the lattice parameter increase and compensate the magnetic moment of Mn atoms. The main element Al does not carry significant magnetic moments, but non-negligible contribution to the magnetic moment comes from conduction electrons.

Antiferromagnetic "ab initio" results were reported for Zr_2MnZ (Z = A, Ga) [26–28] and were gathered in **Table 2**. For Zr_2MnAl, the band gap is slightly increased from 0.41 eV for ferromagnetic calculation to 0.48 eV to antiferromagnetic results. However, the semiconducting band gap from spin-down channel decreases when the atomic radius of the main element increases (when Ga replaces Al). Due to the different magnetic ordering structures, having the spin moments of manganese antiparallel (antiferromagnetic configuration) or parallel (ferromagnetic configuration) oriented, the sign of partial magnetic moments from **Table 2**, differs. However, the opposite spin orientation is clearly explaining the ferrimagnetic

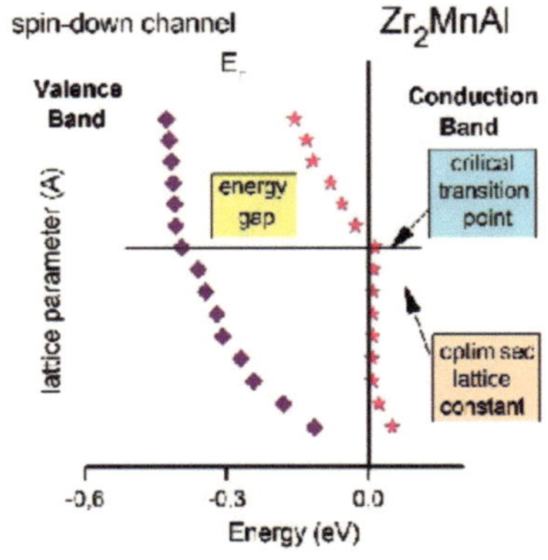

Figure 6.
The positions of the highest occupied states from the valence band (solid rhombs) and of the lowest unoccupied states from the conduction band (solid stars) of total DOSs (spin-down channel) for Zr_2MnAl as function of the lattice parameter.

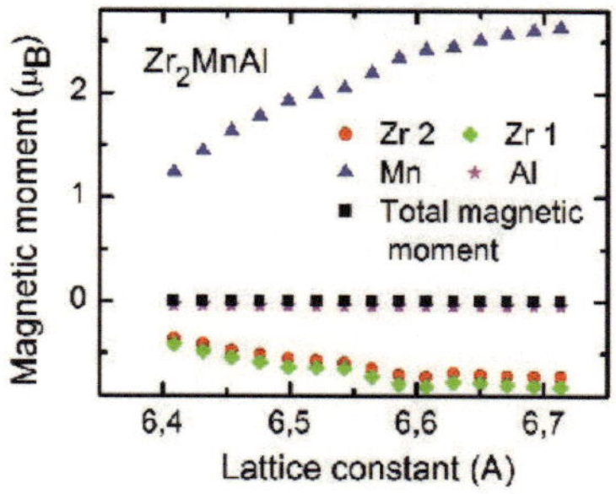

Figure 7.
Total magnetic moments in spin gapless semiconductor Zr2MnAl.

Alloy	a (A)	$\mu_{Zr(4a)}$ (μ_B/atom)	$\mu_{Zr(4c)}$ (μ_B/atom)	$\mu_{Y(4b)}$ (μ_B/atom)	$\mu_{Z(4d)}$ (μ_B/atom)	μ_t (μ_B/f.u.)	E_g (eV)
Zr_2MnAl	6.56[a]	-0.77[a]	-0.69[a]	2.44[a]	-0.04[a]	0.00[a]	0.41[a]
	6.64[b,*]	1.74[b,*]	1.50[b,*]	-3.26[b,*]	0.02[b,*]	0.00[b,*]	
	6.968[d,*]	1.47[d,*]	1.35[d,*]	-4.33[d,*]	0.088[d,*]	0.00[d,*]	0.48[d]
Zr_2MnGa	6.59[c,*]	1.66[c,*]	1.52[c,*]	-3.08[c,*]	-0.10[c,*]	0.00[c,*]	
	6.935[e]	-1.48[e]	-1.42[e]	4.34[e]	-0.030[e]	0.00[e]	0.31[e]

[*] *Antiferromagnetic calculation was performed.*
[a] *Ref. [25].*
[b] *Ref. [26].*
[c] *Ref. [27].*
[d] *Ref. [28].*
[e] *Ref. [29].*

Table 2.
Calculated lattice parameters, partial, total magnetic moments, and energy band gap in Zr_2MnZ (Z = Al, Ga).

interaction between the Mn and Zr atoms and the ferromagnetic coupling between the Zr atoms located in the two distinct sublattice. The total magnetic moment per f.u. calculated in both magnetic configurations is fully compensated by partial magnetic moments of constituents and follow the Slater Pauling curve for typical for half-metals.

4. Half-metallic ferromagnetic materials

Ferromagnetic zirconium-based half-metallic Heusler compounds represent another category of materials of specific interest in biomedical spintronic applications where a good response to an external magnetic moment is required and that is mainly related to their large total magnetic moment. From theoretical point of view, the materials which exhibit a metallic character in the majority density of states and a band gap in the minority one, around the Fermi level and the metallic total density of states resulted from summation of partial density of states of all elements are classified as half-metallic ferromagnets.

A typical example of density of states for a half-metallic ferromagnet is exemplified in case of Zr_2CoAl [30–34] (see **Figure 8**). In the majority channel, the significant contribution to density of states comes from the zirconium, located in the origin of unit cell and the cobalt atom. The band gap from minority channel (**Figure 9**) is formed between the 3d t_{2g} electrons of Co and the 4d t_{2g} unoccupied electrons of Zr located in origin and 4d t_{2g} occupied Zr electrons locate in the 4c Wyckoff position of Hg_2CuTi prototype structure. This type of hybridization between the Y element and the two X atoms of a X_2YZ Heusler compounds is often reported for half-metallic ferromagnets.

In general, the half metallic ferromagnetic properties characterized by the presence of a semiconducting band gap in the minority spin channel of density of states continue to be maintained for a large enough range of lattice parameter in order to be stable from experimental point of view. This means that even if the unit cell volume increases or decreases the material would present a high spin polarization typical for half metallic materials. In case of Zr_2CoAl, the transition from metal behavior to half metallic characteristic occurs at a 6.43 A lattice parameter, and it remains stable at experimentally achievable increases of unit cell volume (**Figure 10**) [30–34]. By definition, in the case of a ferromagnetic material, a net magnetization may be measurable because even if the majority and minority density of states is identical and equally occupied; these are shifted against each other. The total magnetic moment is obtained by adding the partial magnetic moments of constituent elements as presented in **Figure 11**.

Table 3 overviews the state of the art ferromagnetic zirconium-based Heusler compounds. Most alloys incorporate cobalt, and the main elements carry an irrelevant partial magnetic moment. However, the influence of main elements over the energy band gap from spin-down channel is significant. The average width of band gap from spin-down channel is large enough to provide stable half-metallic characteristics for a large deformation of the unit cell. All results

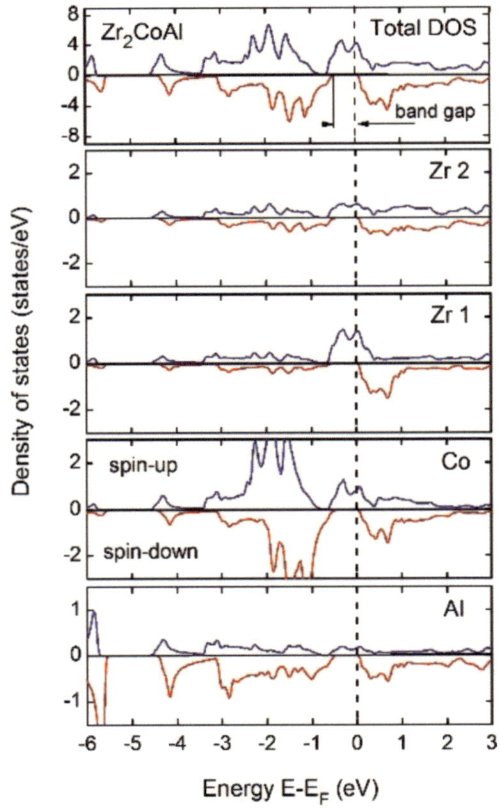

Figure 8.
Partial and total density of states (PDOS, TDOS) of half-metallic ferromagnetic Heusler compound, Zr_2CoAl, at optimized lattice parameter.

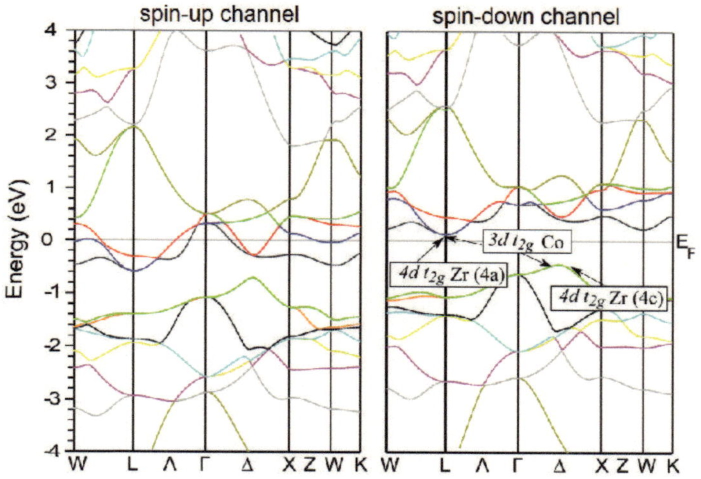

Figure 9.
The band structure of Zr_2CoAl for majority and minority spin channel, in the right and left panel, respectively, at equilibrium lattice parameter.

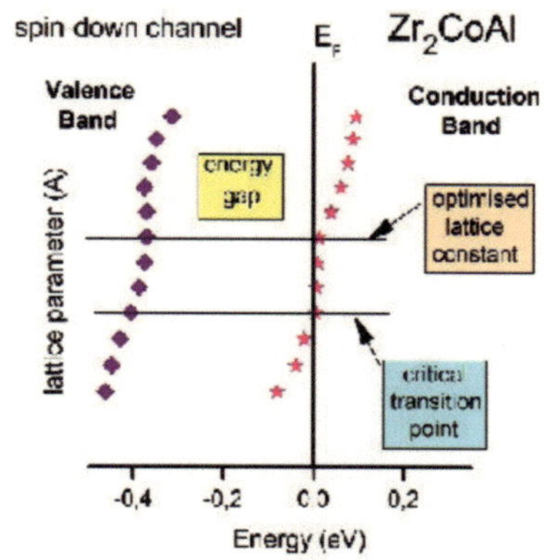

Figure 10.
The positions of highest bonding states from the valence band (solid purple rhombs) and of lowest anti-bonding states from conduction band (solid magenta stars) of total DOSs for Zr_2CoAl as function of lattice constants.

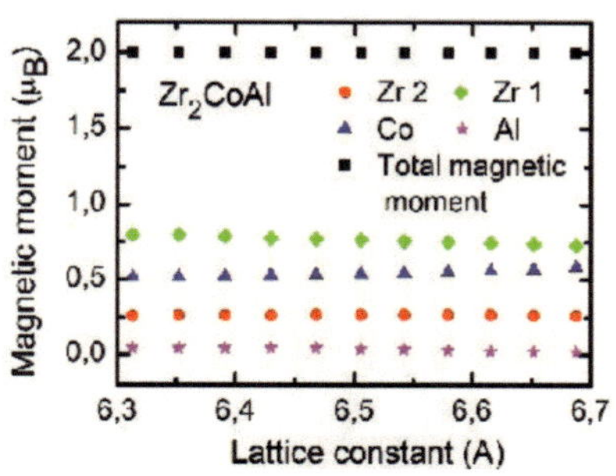

Figure 11.
The partial and total magnetic moments as a function of lattice constant for half-metallic ferromagnetic Zr_2CoAl Heusler compound.

gathered in **Table 3** are obtained based on density functional theory calculations, and these are influenced by the pseudo-potential used for electron-ionic core interaction. This is the reason for various reported band gaps of such compounds, as for example, the band gap reported by Ref [32] which is lower

Alloy	a (Å)	$\mu_{Zr(4a)}$ (μ_B/atom)	$\mu_{Zr(4c)}$ (μ_B/atom)	$\mu_{Y(4b)}$ (μ_B/atom)	$\mu_{Z(4d)}$ (μ_B/atom)	μ_t (μ_B/f.u.)	E_g (eV)
Zr_2CoAl	6.54	0.757[a]		0.54[a]		2.00[a]	0.48[a]
	6.575[i]	1.211[i]	0.303[i]	0.538[i]	-0.053[i]	2.000[i]	0.518[i]
	6.59[c]	1.34[c]	0.36[c]	0.4[c]	-0.1[c]	2.00[c]	0.6046[c]
	6.523[d]	1.088[d]	0.442[d]	0.553[d]	0.002[d]	2.00[d]	0.300[d]
	6.575[e]	1.21[e]	0.3[e]	0.54[e]	-0.05[e]	2.00[e]	0.518[e]
	6.539[f]	0.725[f]	0.262[f]	0.587[f]	0.011[f]	2.00[f]	0.5905[f]
Zr_2CoGa	6.62[c]	1.30[c]	0.52[c]	0.34[c]	-0.16[c]	2.00[c]	0.6990[c]
	6.593[i]	1.162[i]	0.402[i]	0.505[i]	-0.070[i]	2.000[i]	0.533[i]
	6.509[d]	1.074[d]	0.526[d]	0.522[d]	0.013[d]	2.00[d]	0.353[d]
	6.520[f]	0.714[f]	0.332[f]	0.518[f]	-0.001[f]	2.00[f]	0.6546[f]
Zr_2CoIn	6.75[c]	1.34[c]	0.62[c]	0.2[c]	-0.16[c]	2.00[c]	0.7013[c]
	6.627[i]	1.215[i]	0.455[i]	0.417[i]	-0.089[i]	2.000[i]	0.576[i]
	6.714[d]	1.085[d]	0.581[d]	0.427[d]	0.011[d]	2.00[d]	0.268[d]
	6.726[f]	0.722[f]	0.3630[f]	0.429[f]	-0.002[f]	1.999[f]	0.6580[f]
Zr_2CoSi	6.68[c]	1.59[c]	0.58[c]	1.0[c]	-0.08[c]	3.00[c]	0.8419[c]
Zr_2CoGe	6.70[c]	1.65[c]	0.65[c]	0.86[c]	-0.16[c]	3.00[c]	0.8365[c]
Zr_2CoSn	6.76[b]	0.946[b]	0.446[b]	0.8106[b]	-0.013[b]	3.00[b]	0543[b]
	6.790[i]	1.625[i]	0.605[i]	0.829[i]	-0.060[i]	3.000[i]	0.614[i]
	6.81[c]	1.70[c]	0.68[c]	0.78[c]	-0.16[c]	3.00[c]	0.8537[c]
	6.745[d]	1.429[d]	0.746[d]	0.858[d]	0.044[d]	2.998[d]	0.406[d]
	6.745[g]	1.429[g]	0.746[g]	0.858[g]	0.044[g]	3.000[g]	0.65[g]
Zr_2CoPb	6.86[c]	1.72[c]	0.76[c]	-0.16[c]	-0.24[c]	3.00[c]	0.58[c]
Zr_2NiAl	6.60[h]	1.02[h]	0.98[h]	0.61[h]	0.15[h]	2.87[h]	0.44[h]
Zr_2NiGa	6.58[h]	1.06[h]	0.81[h]	0.58[h]	0.21[h]	2.86[h]	0.50[h]

[a]*Ref [30].*
[b]*Ref [35].*
[c]*Ref [31].*
[d]*Ref [32].*
[e]*Ref [33].*
[f]*Ref [34].*
[g]*Ref [36].*
[h]*Ref [37].*
[i]*Ref [21].*

Table 3.
Calculated lattice parameters, partial, total magnetic moments, and energy band gap in Zr_2YZ (Y=Co, Ni; Z = Al, Ga, In, Si, Ge, Sn, Pb).

than the other published results. In all compounds, the ferromagnetic interaction between constituent atoms is represented by similar signs of the partial magnetic moments. The total magnetic moments, following Slater-Pauling curve, are higher than for ferromagnetic half-metallic compounds, and as consequence, the compounds may present a better response to an external magnetic field. Theoretical results regarding zirconium-based half-metallic compounds containing nickel provide information about electronic structures and magnetic properties in alloys with Al and Ga. These intermetallic compounds theoretically behave like half-metals from the electronic structure point of view; however their reported total magnetic moments do not follow the Slater Pauling curve.

In addition, the band gap from spin-down channel is significant lower than for compounds incorporating cobalt.

5. Conclusion

The individualized medicine and high precise diagnosis can benefit from the development of smart biosensors based on magnetic functionalities. Foreseeable applications of zirconium-based biosensors with half metallic character include the capability to measure, sense, or respond to magnetic stimuli desirable for in vivo sensitive detection of markers for diseases.

This chapter overviewed the recent advances of zirconium-based full-Heusler compounds from the point of view of electronic structure and magnetic properties. The representative materials described in this chapter obviously were selected to offer significant information to emphasis the certain differences in magnetic features: half-metallic ferrimagnetism, spin-gapless semiconducting, and half-metallic ferromagnetism. Based on this, the Y elements of Zr_2YZ were selected from the most commonly used transition metals (Cr, Mn, and Co), while the Z element was identical in all compounds (Al). The purpose was to underline the influence of d electrons of Y elements and hybridization interaction between the electrons of zirconium and Y atoms over the macroscopic magnetic properties.

Furthermore, the theoretical and experimental advances in designing and fabrication technology engage the construction of innovative materials to be integrated in biosensors with significant high throughput able to reform the biomedical field.

Acknowledgements

We acknowledge the fruitful discussions with Dr. P. Palade. This work was supported by grants of the Romanian Ministry of Research and Innovation, CCCDI – UEFISCDI, project number PN-III-P1-1.2-PCCDI-2017-0062 contract no 58 and project number PN-III-P1-1.2-PCCDI-2017-0871 contract no 47 as well as by the core program at NIMP.

Author details

Anca Birsan* and Victor Kuncser
National Institute of Materials Physics, Magurele, Romania

*Address all correspondence to: anca_birsan@infim.ro

References

[1] Kübler J, Williams AR, Sommers CB. Formation and coupling of magnetic moments in Heusler alloys. Physical Review B. 1983;**28**:1745

[2] de Groot RA, Mueller FM, van Engen PG, Buschow KHJ. New class of materials: Half-metallic ferromagnets. Physical Review Letters. 1983;**50**:2024

[3] Pickett WE, Moodera JS. Half metallic magnets. Physics Today. 2001;**54**:39

[4] Birsan A, Palade P. Band structure calculations of Ti_2FeSn: A new half-metallic compound. Intermetallics. 2013;**36**:86

[5] de Groot RA. Half-metallic magnetism in the 1990s. Physica B. 1991;**172**:45

[6] Wurmehl S, Fecher GH, Kandpal HC, Felser C. Valence electron rules for prediction of half-metallic compensated-ferrimagnetic behaviour of Heusler compounds with complete spin polarization. Journal of Physics. Condensed Matter. 2006;**18**:6171

[7] Fang Q-L, Zhang J-M, Xu K-W. First-principles investigation of half-metallic ferromagnetism of half-Heusler compounds XYZ. Journal of Magnetism and Magnetic Materials. 2014;**349**:104

[8] Heusler F, Take E. The nature of the Heusler alloys. Transactions of the Faraday Society. 1912;**8**:169

[9] Kandpal HC, Fecher GH, Felser C. Calculated electronic and magnetic properties of the half-metallic, transition metal based Heusler compounds. Journal of Physics D: Applied Physics. 2007;**40**:1507

[10] Slater JC. The ferromagnetism of nickel. Physics Review. 1936;**49**:931

[11] Pauling L. The nature of the interatomic forces in metals. Physics Review. 1938;**54**:899

[12] Kübler J. First principle theory of metallic magnetism. Physica B. 1984;**127**:257

[13] Galanakis I, Dederichs PH, Papanikolaou N. Slater-Pauling behavior and origin of the half-metallicity of the full-Heusler alloys. Physical Review B. 2002;**66**:174429

[14] Skaftouros S, Ozdogan K, Sasioglu E, Galanakis I. Generalized Slater-Pauling rule for the inverse Heusler compounds. Physical Review B. 2013;**87**:024420

[15] Kervan S, Kervan N. Half-metallic properties of the $CuHg_2Ti$-type Mn_2ZnSi full-Heusler compound. Current Applied Physics. 2013;**13**:80

[16] Bayar E, Kervan N, Kervan S. Half-metallic ferrimagnetism in the Ti_2CoAl Heusler compound. Journal of Magnetism and Magnetic Materials. 2011;**323**:2945

[17] Birsan A. Electronic structure and magnetism of new scandium-based full Heusler compounds: Sc_2CoZ (Z = Si, Ge, Sn). Journal of Alloys and Compounds. 2014;**598**:230

[18] Ahmadian F, Salary A. Half-metallicity in the Inverse Heusler compounds Sc_2MnZ. Intermetallics. 2014;**46**:243

[19] Zheng N, Jin Y. Band-gap and Slater-Pauling rule in half-metallic Ti_2-based Heusler alloys: A first-principles study. Journal of Magnetism and Magnetic Materials. 2012;**324**:3099

[20] Ouardi S, Fecher GH, Felser C, Kübler J. Realization of spin gapless semiconductors: The Heusler compound Mn_2CoAl. Physical Review Letters. 2013;**110**:100401

[21] Wei X-P, Zhang Y-L, Wang T, Sun X-W, Song T, Guo P. Stability,

electronic and magnetic properties investigations on Zr_2YZ (Y = Co, Cr, V and Z = Al, Ga, In, Pb, Sn, Tl) compounds. Materials Research Bulletin. 2017;**86**:139-145

[22] Deng Z-Y, Zhang J-M. Half-metallic and magnetic properties of full-Heusler alloys Zr_2CrZ (Z=Ga, In)with Hg_2CuTi-type structure: A first-principles study. Journal of Magnetism and Magnetic Materials. 2016;**397**:120-124

[23] Li HZ, Gao Y-C, Wang XT, Wang L-Y. Prediction of the half-metallic properties of Zr_2VZ (Z = Si, Ge, Sn, and Pb) Heusler alloys based on density functional theory. Journal of Superconductivity and Novel Magnetism. 2016;**29**:493-500

[24] Gao YC, Wang XT, Rozale H, Lu JW. Theoretical investigations of electronic structures, magnetic properties and half-metallicity in Heusler alloys Zr2VZ (Z = Al, Ga, In). Journal of the Korean Physical Society. 2015;**67**:881-888

[25] Birsan A, Kuncser V. First principle investigations of the structural, electronic and magnetic properties of predicted new zirconium based full-Heusler compounds, Zr_2MnZ (Z=Al, Ga and In). Journal of Magnetism and Magnetic Materials. 2016;**406**:282-288

[26] Wang XT, Cheng ZX, Wang JL, Rozale H, Wang LY, Yu ZY. Strain-induced diverse transitions in physical nature in the newly designed inverse Heusler alloy Zr_2MnAl. Journal of Alloys and Compounds. 2016;**686**:549-555

[27] Wang X, Cheng Z, Khenata R, Wu Y, Wang L, Liu G. Lattice constant changes leading to significant changes of the spin gapless features and physical nature in a inverse Heusler compound Zr_2MnGa. Journal of Magnetism and Magnetic Materials. 2017;**444**:313-318

[28] Patel PD, Shinde SM, Gupta SD, Jha PK. A promising thermoelectric response of fully compensated ferrimagnetic spin gapless semiconducting Heusler alloy Zr_2MnAl at high temperature: DFT study. Materials Research Express. 2019;**6**:076307

[29] Patel PD, Shinde S, Gupta SD, Dabhi SD, Prafulla K. The first principle calculation of structural, electronic, magnetic, elastic, thermal and lattice dynamical properties of fully compensated ferrimagnetic spin-gapless Heusler alloy Zr_2MnGa. Computational Condensed Matter. 2018;**15**:61-68

[30] Birsan A. Magnetism in the new full-Heusler compound, Zr_2CoAl: A first-principles study. Applied Physics. 2014;**14**:1434-1436

[31] Wang XT, Cui YT, Liu XF, Liu GD. Electronic structures and magnetism in the Li_2AgSb-type Heusler alloys, Zr_2CoZ (Z=Al, Ga, In, Si, Ge, Sn, Pb, Sb): A first-principles study. Journal of Magnetism and Magnetic Materials. 2015;**394**:50-59

[32] Yan P-L, Jian-Min-Zhang, Xu K-W. Electronic structures, magnetic properties and half-metallicity in Heusler alloys Zr_2CoZ (Z=Al, Ga, In, Sn). Journal of Magnetism and Magnetic Materials. 2015;**391**:43-48

[33] Wei X-P, Sun W, Zhang Y-L, Sun X-W, Song T, Wang T, et al. Investigations on electronic, Fermi surface, Curie temperature and optical properties of Zr_2CoAl. Journal of Solid State Chemistry. 2017;**247**:97-104

[34] Amari D, Mokhtari M, Dahmane F, Belfarh T, Tabeti A, Elkeurti M, et al. A comparative study between Hg_2CuTi and Cu_2MnAl type structures for Zr_2CoZ (Z = Al, Ga, In) Heusler alloys. Chinese Journal of Physics. 2019;**60**:450-461

[35] Birsan A. Theoretical investigations of electronic structure and magnetism in Zr_2CoSn full-Heusler compound.

Journal of Magnetism and Magnetic Materials. 2015;**388**:1-4

[36] Yang Y, Feng Z-Y, Zhang J-M. Surface thermodynamic stability, electronic and magnetic properties in various (001) surfaces of Zr_2CoSn Heusler alloy. Superlattices and Microstructures. 2018;**117**:82-91

[37] Yousuf S, Gupta DC. Investigation of electronic, magnetic and thermoelectric properties of Zr_2NiZ (Z = Al,Ga) ferromagnets. Materials Chemistry and Physics. 2017;**192**:33-40

Section 3

Dilute Magnetic Semiconductor

Chapter 6

Diluted Magnetic Semiconductor ZnO: Magnetic Ordering with Transition Metal and Rare Earth Ions

Kuldeep Chand Verma

Abstract

For advancement in future spintronics, the diluted magnetic semiconductors (DMSs) might be understood for their origin of ferromagnetic aptness. It not much clear to the ferromagnetism in DMS, that is intrinsic or via dopant clustering formation. For this, we have included a review study for the doping of transition metal and rare earth ions in ZnO. It is realized that the antiferromagnetic ordering is found in doped ZnO to achieve high-T_C ferromagnetism. X-ray diffraction and Raman spectra techniques have been used to detect the wurtzite ZnO structure and lattice defects. Since ZnO has different types of morphology formation that is generally dependent on synthesis conditions and dopant level. The band gap energy of ZnO and lattice defect formation are shown by photoluminescence technique. The room temperature ferromagnetism is described with bound magnetic polaron (BMP) model in which oxygen vacancies play a major role. However, the temperature-dependent conditions are responsible for ferromagnetic ordering. The first principle calculation is used for dopant ions in ZnO for their replacement of Zn^{2+} atoms in the wurtzite structure as well as magnetic contribution.

Keywords: ZnO ferromagnetism, lattice structure, TEM, vacancies

1. Introduction

Diluted magnetic semiconductors (DMSs) and oxides raised hopes of applications based on single elements combining the logic functionalities of semiconductors with the information storage capabilities of magnetic elements [1]. The research on DMSs is fascinating during the 1980s on II-VI semiconductors with little magnetic impurities that opened the optical and electronic phenomenon to introduce magnetic field effects [2]. The DMS required high Curie temperature, T_C, for a material. Initially, II-VI semiconductors such as CdMnTe and ZnMnSe are widely investigated as DMSs for which the valence of the cation matches with the common magnetic ions, i.e., Mn [3]. Since the hole induced ferromagnetic order in p-type InMnAs and GaMnAs, a lot of research is carried out on III-V-based DMSs that unfortunately failed to give room temperature ferromagnetism (RTFM). After that, a lot of efforts are carried out to result RTFM in III-V-based DMS, i.e., GaN, GaSb, InAs, and oxide-based DMS, ZnO, TiO_2, SnO_2, In_2O_3, etc. These DMSs are attracting potential interest in spin-based information-processing applications. It needs high T_C for such DMSs of GaN and ZnO that may relate for their wide-band gap [4]. However, the spintronic applications like spin-valve transistors, spin light-emitting diodes, nonvolatile memory, logic devices, etc. have remarkable interest of RTFM of DMSs [5]. Among DMSs, the Mn-doped GaAs is found to be ferromagnetic with $T_C \sim 172$ K is widely investigated [5].

1.1 DMS opens new window for spintronics

Even though low transition temperatures, ferromagnetism in diluted magnetic semiconductors, DMSs is essential to explore new ideas to develop spintronic technology, which is actually the electrical manipulation of magnetism [6]. A small Mn percentage in III-V semiconductors of (In, Mn)As and (Ga, Mn)As replaces Ga or In atoms to stabilize the exchange interaction between charge carriers and localized spins. Due to low carrier concentration in these DMSs, it is possible to control a considerable portion of carriers by external electric fields using metal–insulator–semiconductor or p-n junction configuration (**Figures 1a, b**). It has a thin ferromagnetic semiconductor layer, to the extent that the field significantly alters the stability of the ferromagnetic phase and other magnetic properties. The value of T_C is controlled in a ferromagnet with the application of an electric field which was observed for metal–insulator–semiconductor configuration of (In, Mn)As thin films [8]. Later, electrical manipulation of the coercive field (H_C) is also possible for (In, Mn)As which means an applied electric field changes the magnetic anisotropy [9]. This is the exchange interaction which splits the carrier states according to the spin-orbit interaction [10].

1.2 DMS made up as a computer memory

For low-power-consumption computer memory devices, the DMSs influencing the magnetization direction to achieve magnetic data-storage and memory devices of hard disks directs with the direction of magnetization [7]. In a hard disk, the data is stored on a disk-shaped magnet in local magnetization form; to write information, a pulse of current is applied to a small electromagnet that scans the disk. In this process, an energy is wasted due to a magnetic field exists between the current and the manipulated magnetization. However, a spin-polarized current is applied directly to the magnet instead of using current to generate the magnetic field is an

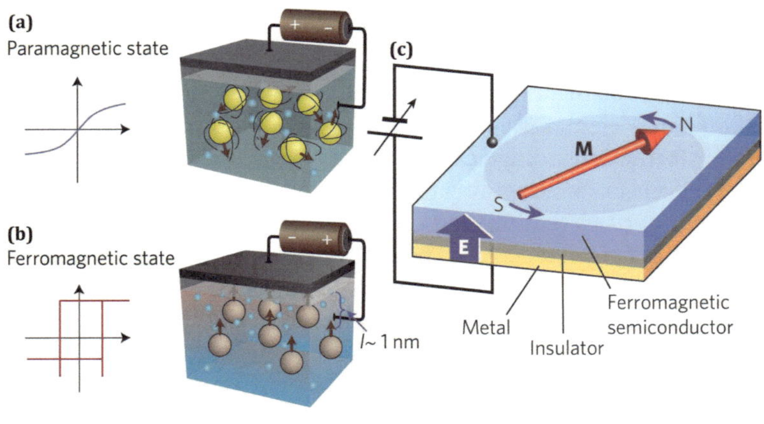

Figure 1.
(a, b) Electric control of ferromagnetism (schematic representation). (c) a spin-polarized current changes the magnetization direction, i.e., (GaMn)As (adapted from [6, 7]).

alternative way as schematized in **Figure 1c**. Such a current would exert torque on the magnetization by exchanging spin-angular momentum with it as it passes through the magnet. Exploitation of this phenomenon, called spin torque, is expected to allow the development of compact magnetic memory devices that can run on low-power consumption. In **Figure 1c**, the metal–insulator–semiconductor device involving a semiconductor—a (GaMn)As film—that has ferro magnetic properties at low temperatures [7]. The device includes a 'gate' electrode isolated electrically from the (GaMn)As film. When a negative voltage is applied to the gate electrode, carriers in the film that have positive charge (electron "holes") are attracted toward the electrode and vice versa. This property allows the density of the electron holes, and thus the magnetic anisotropy in the (GaMn)As film beneath the electrode, to be controlled electrically, resulting in a change in the magnetization direction.

1.3 Ferromagnetic origin in DMS

The researcher has initially found high T_C in doped III-V DMS, which for a long time was stuck at 110 K [1]. After that, several groups stressed out the defects—mainly Mn atoms that form interstitials rather than substituting for Ga—responsible for this limit, and T_C was raised up to 150 K. Dietl [1] proposed a Zener model to perform so many experiments on (Ga, Mn)As, which create problem with higher concentration of Mn due to the interplay between the disorder and localization, and electron–electron correlations have a very influential effect on carrier-mediated ferromagnetism at and above room temperature [11]. For example, for the Co-doped ZnO, the Co occupies the Zn sites as Co is paramagnetic and there is no ferromagnetism associated with Co, even when lots of carriers are added by Al co-doping and the temperature is very low (5 K) [12]. Moreover, the solubility of Co in ZnO is high, making it is easy to substitute Co for Zn throughout the crystal. However, by considering defect-mediated ferromagnetism, an intrinsic form of high-T_C ferromagnetism in dilute magnetic oxides with lots of defects is observed. This is because electrons associated with defects couple antiparallel to dopant spins within the orbital volume of the defect. With high T_C, the defect coupling is strong. But, it is difficult to control defects for practical applicability. To realize high-temperature ferromagnetism in DMSs, a wide-band gap ZnO is undoubtedly a major development if the ferromagnetism is unambiguously established to be intrinsic (carrier induced) [13]. Coey et al. [14] proposed that the ferromagnetic exchange is mediated by shallow donor electrons to form bound magnetic polarons that overlap to create a spin-split impurity band. It is reported that the oxygen vacancies might change the band structure of host oxides to induce ferromagnetism [15]. The formation of BMP, which includes electrons locally trapped via oxygen vacancies, with the trapped electron occupying an orbital overlapping with the d shells of transition metal (TM) neighbors, might explain the room temperature ferromagnetism (RTFM) in DMS. Within the BMP model, the greater density of oxygen vacancy yields a greater overall volume occupied by BMP, thus increasing their probability of overlapping more TM ions into the ferromagnetic domains to enhance ferromagnetism. Zhen et al. [16] used first principle calculations on Co-doped ZnO and observed the exchange coupling mechanism that accounts magnetism with oxygen vacancies.

1.4 DMS ZnO

DMS ZnO has the hexagonal wurtzite structure (direct wide-band gap, $E_g \sim 3.3$ eV at 300 K) due to its stability at room temperature and normal atmospheric pressure. The atomic arrangement of the wurtzite structure is comprised of four zinc ions (Zn^{2+}) occupying the corner of a tetrahedral coordinate with one oxygen ion (O^{2-}) located at the center and vice versa (**Figure 2a**) [20]. The particle size, doping, and co-doping are used to induce the band gap of ZnO [17]. Dietl et al. [1] reported RTFM for DMS for which 3d ions substituted ZnO. However, the nanocrystals of pure ZnO also produce RTFM [21]. Gao et al. [22] suggested oxygen vacancies locating at the surface of ZnO nanoparticles are responsible for RTFM. It is also found that the ZnO nanoparticles had absorbed certain organic molecules to modify the electronic structure to give RTFM without any magnetic impurity ions [21].

1.4.1 Ferromagnetism of ZnO with transition metal ions

From the survey of many theoretical studies, it has been found that a slight doping of TM metal ions is likely ∼5%, induce ferromagnetic ordering that observed at room temperature [21]. Venkatesan et al. [18] postulated on the basis of spin-split donor impurity-band model to observe RTFM in DMS ZnO with 5% of Sc, Ti, V, Fe, Co, or Ni, but not Cr, Mn, or Cu ions. For Cr, Mn, Cu, or Zn, no moment appreciably greater than the experimental uncertainty ($<0.1\ \mu_B$) is observed at room temperature. The basic action in a spintronic device is that the electrons are traveling from a ferromagnetic metal, through a normal metal, to a second ferromagnetic metal. When the magnetizations of the two ferromagnetic metals are in an aligned state, the resistance is low, whereas the resistance is high in the antialigned state. For the light 3d elements, the $3d^{\uparrow}$ states lie high in the $2p$(O)-4s(Zn) gap, overlapping the donor impurity band which is spin split (**Figure 2b**). In the middle of the TM series, there is no overlap with the 3d levels and exchange is weak, but toward the end of the series, the $3d^{\downarrow}$ states overlap the impurity band, which then has the opposite spin splitting for the same occupancy. The high T_C is found whenever unoccupied $3d$ states overlap the impurity band, but not otherwise. The likely origin of the donor impurity band in ZnO films is lattice defects, such as oxygen vacancies, which have trapped between one and two electrons (F^0 centers)

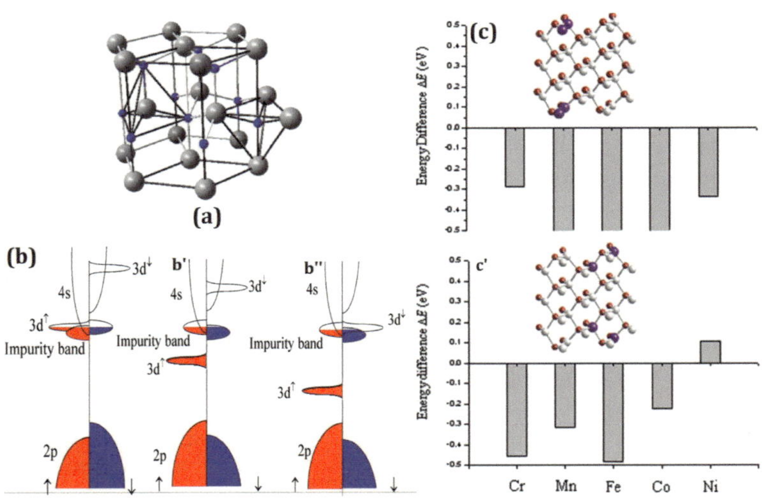

Figure 2.
(a) Hexagonal wurtzite ZnO unit cell. Density of states (schematic) of $Zn_{1-x}TM_xO$, TM = Ti (b), Mn (b′), Co (b″), for which the Fermi level lies in a spin-split donor impurity band. (c) Energy differences, $\Delta E = E_{AFM} - E_{FM}$, for $Zn_{0.857}TM_{0.143}O$ (TM = Cr, Mn, Fe, Co, and Ni) supercell; Zn (light gray spheres), O (red), and TM (purple) (adopted from [17–19]).

[18]. The electrons in the impurity band will be localized by the influence of electronic correlations and potential fluctuations associated with the dopant cations.

1.4.2 Theoretical survey on magnetism of DMS ZnO with TM = Cr, Mn, Fe, Co, and Ni ions

Wang et al. [19] reported that the Cr, Fe, Co, and Ni dopants in ZnO occupy the Zn sites and couple antiferromagnetically, while Mn exhibits no site preference and distributes uniformly in ZnO lattice. For hexagonal ZnO, the lattice constants are $a = b = 3.249$ Å and $c = 5.205$ Å [space-group $P6_3mc$ (No. 186)]. The ZnO thin film containing 28 formula units ($Zn_{28}O_{28}$) is shown in **Figure 1c** [19]. To find the magnetic coupling among TM ions, the two Zn atoms are replaced with two TM (= Cr, Fe, Co, and Ni) atoms with a dopant concentration of 14.28%. The preferred magnetic coupling between the TM atoms is determined with ferromagnetic (FM) and antiferromagnetic (AF) coupling by comparing their total energies, $\Delta E = E_{AF} - E_{FM}$. Positive ΔE means that the FM state is lower in energy than the AF state. In **Figure 1c**, when the two TM atoms are at the nearest neighbor sites on the surface, the corresponding magnetic couplings are AF. It is also reported that the total-energy difference between FM and AF states is reduced to 0.006–0.032 eV, when the distance between two TM atoms are increased to about 5.60 Å. It means that the AF interactions are short ranged in TM/ZnO. Srinivasulu et al. [23] suggested various 3d TM such as Ti, V, Cr, Mn, Fe, Co, Ni, and Cu that are also tried as dopants in ZnO to improve its optical and electrical behavior. Among these dopants, V, Cr, Mn, Co, Ni, and Cu are recognized as suitable dopants of ZnO for spintronic and magneto-optical communication devices due to their RTFM. In TM ions, the magnetization arises from partially filled 3d shells, and most of the cases since total orbital magnetic moment is zero, the magnetic moment is only due to the spin component, and hence total magnetic moment per atom is less [24]. Among TM/ZnO, Co deserved a special attention due to its highest magnetic moments (4.8 μ_B) and a positive magnetic exchange coupling constant [25]. Coey et al. [14] explained ferromagnetism in intrinsically n-type semiconductors and insulators by a model, where shallow donor electrons, created due to intrinsic defects in the semiconductors, form bound magnetic polarons with magnetic cations, which finally give rise to the ferromagnetic interaction. For BMPs, the localized spins of the dopant ion interact with the charge carriers such as oxygen vacancies, resulting in a magnetic polarization of the surrounding local moments [26]. The mediated oxygen vacancies are dependent upon dopant level and nanostructural formations.

1.4.3 Rare earth ions attributed ferromagnetism in DMS ZnO

In rare earth (RE) elements, magnetization appears due to unfilled 4f orbitals leading to higher magnetic moment per atom, though 4f electrons interacted with 5d or 6 s electrons [24], but exhibits weak exchange interaction with other RE ions, which is contrast to TM ions 3d electrons are directly interacted. The RE ion-doped ZnO has ferromagnetism that is induced by p-f hybridization via defect carriers [27]. Compared with 3d TMs, 4f RE ions have larger magnetic moments. The intrinsic defects such as oxygen vacancies play an important role on the magnetic properties of RE/ZnO. However, the exchange interaction by simultaneous doping from TM and RE ions in ZnO is 4f-5d-3d, which is antiferromagnetic when the 5d band is less than half full and the 3d band is more than half full. The first principle calculations revealed that the superexchange interaction between two magnetic Nd ions is mediated by the nonmagnetic O ions responsible for higher magnetic moment of ZnO [28]. This approach of doping RE elements with intrinsic strong magnetic anisotropy and tailoring the coupling between dopants and defects should be a general approach toward stable ferromagnetic order in ZnO nanomaterials. Among RE ions, Sm^{3+} with five 4f electrons offers a unique possibility to induce the bifunctional properties for RTFM as well as visible luminescence in ZnO, making suitable material in spin transport properties and spin-LEDs [29].

2. Experimental methods

The DMS ZnO materials are synthesized by different methods such as thermal evaporation method [30], chemical vapor deposition [31], sol-gel spin-coating technique [32], spray pyrolysis [33], hydrothermal synthesis [34], solid state reaction [29], coprecipitation method [35], etc.

3. Results and discussion

Recently, a lot of research work has been reported on RE ion-based DMS ZnO [36]. Sun et al. [37] reported La-doped ZnO quantum dots in which luminescent behavior is greatly enhanced by introducing defects and oxygen vacancies (V_O). This is due to larger ionic size of La^{3+} in ZnO lattice that induces stress. However, Bantounas et al. [38] suggested the weak magnetic coupling in Gd/ZnO and the material remain paramagnetic at room temperature. Aravindh et al. [39] gives origin of ferromagnetism in Gd/ZnO in which oxygen vacancies play an important role. Using DFT calculation, it is analyzed that the RE Ce atoms replaced those Zn sites in the wurtzite structure, which is the nearest neighbor to TM/Fe or Co atoms [40]. The 4f electrons in Ce are tightly bound around the nucleus and shielded by $5s^2p^6d^16s^2$ electrons, leading to strong local spin. For Ce-doped TM/ZnO, the larger ratio of dopant cation to cation radius structure causes more defects, leading to a larger concentration of electrons and holes.

3.1 Wurtzite structure and defect calculation in DMS ZnO

3.1.1 X-ray diffraction of $Zn_{0.94}Fe_{0.03}Ce_{0.03}O$ and $Zn_{0.94}Co_{0.03}Ce_{0.03}O$ nanoparticles

The $Zn_{0.94}Fe_{0.03}Ce_{0.03}O$ (ZFCeO) and $Zn_{0.94}Co_{0.03}Ce_{0.03}O$ (ZCCeO) nanoparticles were synthesized by a sol-gel process [40]. **Figure 3a** shows the X-ray diffraction (XRD) results for ZFCeO and ZCCeO nanoparticles using Rietveld refinement (space group $P6_3mc$). The Rietveld refinement initiated with Zn^{2+} and O^{2-} atoms is located at (1/3, 2/3, 0) and (1/3, 2/3, z), respectively. The XRD reflections result into a hexagonal wurtzite ZnO phase. The refined lattice parameters are a(Å) = 3.259(1) and 3.262(3) and c(Å) = 5.215(3) and 5.218(2); unit cell volume, V(Å^3) = 47.9682(3) and 48.0828(2); bond length, $l_{Zn\text{-}O}$(Å) = 1.9826 and 1.9842; R_p(%) = 6.57 and 6.95; R_{wp}(%) = 9.0 and 9.8; and χ^2 = 1.97 and 2.05, respectively, for ZFCeO and ZCCeO. Lattice parameters for the hexagonal wurtzite ZnO structure is also calculated using the relation

$$\frac{1}{d^2} = \frac{4}{3}\frac{\left(h^2 + hk + k^2\right)}{a^2} + \frac{l^2}{c^2} \tag{1}$$

where a, c, h, k, l, and d have their usual meaning. The value of bond length is calculated [40]:

$$l_{Zn-O} = \sqrt{\frac{a^2}{3} + \left(\frac{1}{2} - u\right)^2 * c^2} \tag{2}$$

where

$$u = \frac{a^2}{3c^2} + 0.25 \tag{3}$$

where u is a positional parameter. The volume per unit cell for the hexagonal system is calculated using.

$$V = 0.866 \times a^2 \times c \tag{4}$$

The calculated values of the lattice parameters are a(Å) = 3.257, 3.256, 3.260, and 3.261; c(Å) = 5.207, 5.206, 5.214, and 5.217; c/a = 1.5987, 1.5988, 1.5994, and 1.5998;

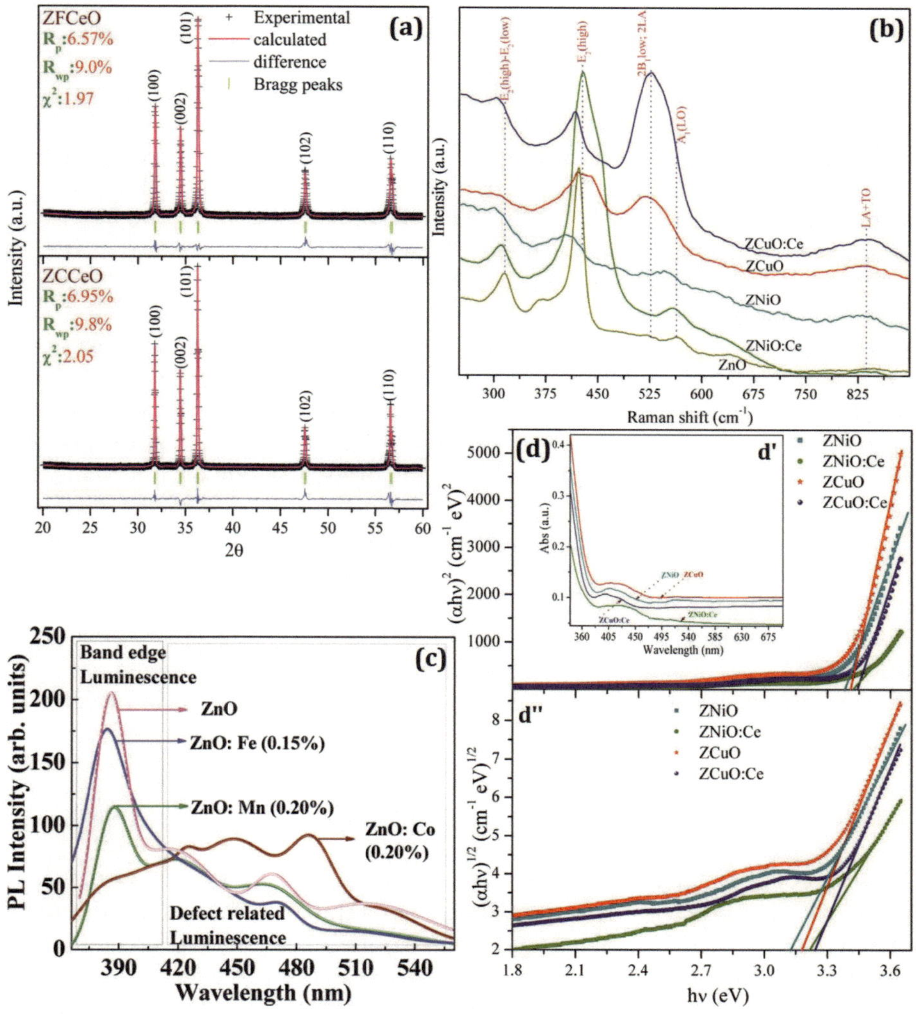

Figure 3.
(a) XRD pattern of $Zn_{0.94}Fe_{0.03}Ce_{0.03}O$ (ZFCeO) and $Zn_{0.94}Co_{0.03}Ce_{0.03}O$ (ZCCeO) nanoparticles. (b, d) Raman and UV-visible absorption (for energy band calculation) spectra of $Zn_{0.95}Ni_{0.05}O$ (ZNiO), $Zn_{0.91}Ni_{0.05}Ce_{0.04}O$ (ZNiO/Ce), $Zn_{0.95}Cu_{0.05}O$ (ZCuO), and $Zn_{0.91}Cu_{0.05}Ce_{0.04}O$ (ZCuO/Ce) nanoparticles. (c) Photoluminescence spectra for ZnO with Co, Mn, and Fe nanoparticles (adapted from [27, 40, 41]).

$l_{Zn\text{-}O}$(Å) = 1.9808, 1.9825, 1.9829, and 1.9837; and V(Å^3) = 47.834, 47.796, 47.987, and 48.044, calculated for $Zn_{0.97}Fe_{0.03}O$ (ZFO), $Zn_{0.97}Co_{0.03}O$ (ZCO), ZFCeO, and ZCCeO, respectively. The calculated values of the c/a ratio of ZFO and ZCO are slightly increased over pure ZnO (c/a = 1.598) due to the shape/size effect of the nanorods. But, it is again enhanced with RE ions due to the ionic size effect. Therefore, the observed variation in lattice parameters with doping indicated displacement of atoms in wurtzite lattice to create defects, i.e., vacancies or interstitials. It is also reported that the average size, D, of nanoparticles is 97 ± 4 nm and 106 ± 3 nm for ZFCeO and ZCCeO, respectively. The lattice defects are also evaluated with Raman and photoluminescence spectra. The zero-field cooling (ZFC) and field cooling (FC) magnetization measurement at H = 500 Oe and T = 300–5 K show AF-FM transitions. At 5 K, the measured value of M_s(emu g^{-1}) = 0.339 and 0.478 for ZFCeO and ZCCeO, respectively. For ZFCeO, the weak RTFM is formed due to the mixed valance states Fe^{2+}/Fe^{3+} via oxygen vacancies.

3.1.2 Lattice structure and defect/vacancy evaluation by Raman spectra

The $Zn_{0.95}Ni_{0.05}O$ (ZNiO), $Zn_{0.91}Ni_{0.05}Ce_{0.04}O$ (ZNiO/Ce), $Zn_{0.95}Cu_{0.05}O$ (ZCuO), and $Zn_{0.91}Cu_{0.05}Ce_{0.04}O$ (ZCuO/Ce) nanoparticles were synthesized by sol-gel process [27]. XRD pattern found wurtzite structure with lattice distortion to perform lattice defects. The average particles size is D = 27, 81, 57 and 159 nm, respectively, measured for ZNiO, ZNiO/Ce, ZCuO, and ZCuO/Ce. The Raman modes observed at room temperature for these pure Ni-, Cu-, and Ce-doped ZnO are shown in **Figure 3b**. The presence of E_2 mode in all samples indicates that the doping does not change the wurtzite phase. It is observed that Ni and Cu doping on ZnO gradually decreases the intensity of E_2(high) mode as compared with pure ZnO [22]. But, it is again strengthen with Ce co-doping. This type of change in E_2(high) mode with dopant ions might induce structural defects and local lattice distortions of wurtzite lattice [32]. The E_2(high)-E_2(low) modes indicate oxygen defects or vacancy formation. The peak position of E_2(high) mode also changes with Ni, Cu, and Ce doping that is ascribed with the change in the level of oxygen vacancies [33]. The magnetic results also reported low temperature ZFC/FC magnetic measurement that show AF-FM ordering and the doping of Ce ions results to high T_c. At 300 K, the values of M_s(emu g^{-1}) = 0.073, 0.085, 0.053, and 0.132, and at 10 K M_s(emu g^{-1}) = 0.096, 0.198, 0.136, and 0.251, respectively, for ZNiO, ZNiO/Ce, ZCuO, and ZCuO/Ce. The enhancement in the oxygen vacancies and ferromagnetism with Ce doping might depend on mixed valence state Ce^{3+}/Ce^{4+} ions.

3.1.3 Photoluminescence spectra for Fe-, Co-, and Mn-doped ZnO nanoparticles

The photoluminescence spectra for Fe (0.15%)-, Co (0.20%)-, and Mn (0.20%)-doped ZnO nanoparticles are given in **Figure 3c** [41]. Pure ZnO nanoparticles show emission maxima at 385 nm along with blue (424 nm, 468 nm) and green (521 nm) luminescence. Transitions from Zn interstitials to valence band are attributed with blue emission (424 nm). Oxygen vacancies are related with blue (468 nm) and green emission (521 nm). The green emission is understood to be due to the recombination of electrons in singly occupied oxygen vacancies with photoexcited holes in the valence band. The blue emission is caused by two defect levels, either transition from Zn_i to the valance band or transition from bottom of the conduction band to the interstitial O(O_i).

3.1.4 UV-Visible absorption spectra and Tauc plot

Figure 3d′ shows UV-visible absorption spectra measured at room temperature for Ni-, Cu-, and Ce-substituted ZnO nanoparticles. The absorption peaks are observed corresponding to violet emission, i.e., ZNiO (409 and 426 nm), ZNiO/Ce (433 nm), ZCuO (407 and 427 nm), and ZCuO/Ce (401 and 429 nm). In order to evaluate the effect of dopant on Ni, Cu, Ce, on ZnO, the energy band gap, Eg, is calculated using the Tauc relation [27] used for a direct transition using: $(\alpha h\nu)^2 = K(h\nu - E_g)$; and for an indirect transition:$(\alpha h\nu)^{1/2} = K(h\nu - E_g)$. The symbols in these equations have their usual meanings. In **Figure 3d**, the value of direct band energy, E_g(direct) = 3.38, 3.42, 3.41, and 3.44 eV, and from **Figure 7d″**, the indirect energy band gap, E_g(indirect) = 3.13, 3.21, 3.19 and 3.24 eV, respectively measured, for ZNiO, ZNiO/Ce, ZCuO, and ZCuO/Ce. These values of E_g show small variation with bulk sample of pure ZnO [27]. However, a significant change in E_g value from direct and indirect measurement clearly indicates that some of the defect states are present in the forbidden region.

3.2 Microstructural study of DMS ZnO

3.2.1 SEM image of Mn-doped ZnO nanowires

The ZnO nanowires were synthesized by a thermal evaporation method with 1 atom % Mn doping [44]. **Figure 4a** is a SEM image of Mn/ZnO nanowires of several micrometer lengths and 70 nm diameters. The reported work given T_C to be 437 K

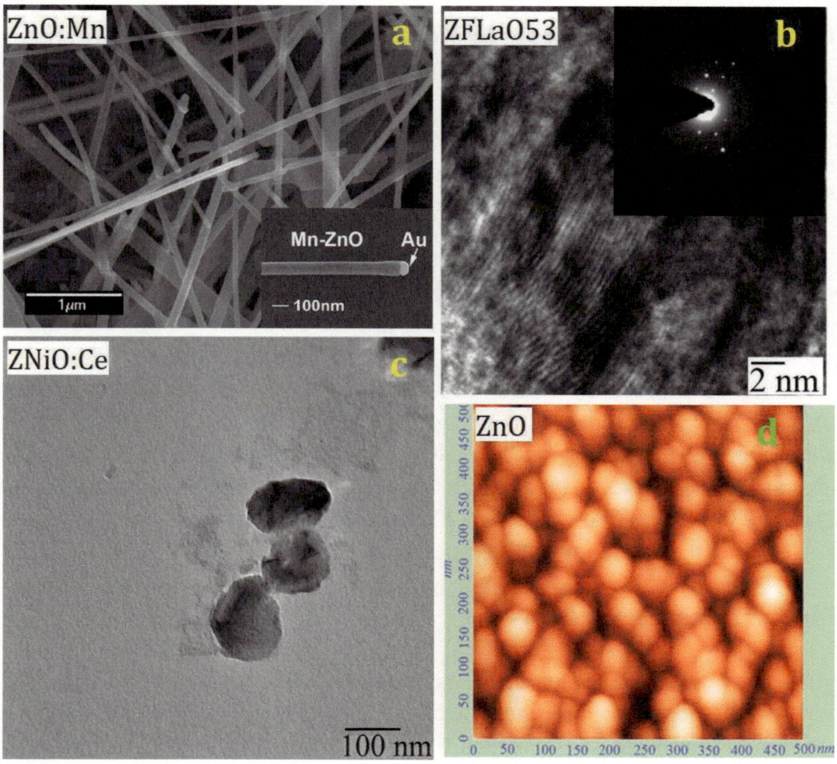

Figure 4.
(a) Scanning electron microscopy (SEM) pattern for ZnO/Mn nanowires. (b) HRTEM for $Zn_{0.92}Fe_{0.05}La_{0.03}O$ nanoparticles. (c) TEM pattern for $Zn_{0.91}Ni_{0.05}Ce_{0.04}O$ nanoparticles. (d) AFM pattern for pure ZnO (adapted from [21, 27, 44, 45]).

from SQUID measurements. The difference of ΔMR is up to 2.5% as the gate voltage changes from −40 to +40 V at T = 1.9 K, which suggests the electric field control of ferromagnetism for realizing spin logic devices.

3.2.2 HRTEM of $Zn_{0.92}Fe_{0.05}La_{0.03}O$ nanoparticles

DMSs $Zn_{0.92}Fe_{0.05}La_{0.03}O$ (ZFLaO53) nanoparticles were synthesized by sol-gel process [21]. The value of nanoparticles size is 99 nm. The lattice spacing is calculated from high-resolution transmission electron microscopy (HRTEM) images (**Figure 4b**), which show that the distorted lattice has an enhanced interplanar spacing d [(101) planes] of ~ 0.247 nm. It is also observed from HRTEM that some fractions within the lattice fringes are formed. This may due to some ferromagnetic clustered growth by dopants in Zn^{2+} lattice. The high crystallinity of the particles is evident from the selected area electron diffraction.

3.2.3 TEM of $Zn_{0.91}Ni_{0.05}Ce_{0.04}O$ nanoparticles

The $Zn_{0.91}Ni_{0.05}Ce_{0.04}O$ (ZNiO/Ce) nanoparticles were synthesized by a sol-gel process [27]. **Figure 4c** shows their TEM image with an average size of nanoparticles of 81 nm.

3.2.4 Atomic force microscopy (AFM) of pure ZnO

The ZnO thin film is prepared by a sol–gel MOD method [45] with the average size of nanoparticles of 40 nm (**Figure 4d**).

3.3 First principle calculation for DMS ZnO with TM and RE ions

3.3.1 Magnetic behavior of $Zn_{1-x}TM_xO$ (T = Cr, Mn, Fe, Co, and Ni)

The magnetic properties of $Zn_{1-x}TM_xO$ (T = Cr, Mn, Fe, Co, and Ni) thin films are investigated using first principle calculations on the basis of DFT theory within the generalized gradient approximation (GGA) [19]. Self-consistency is achieved by allowing the total energy to converge within 1 meV because of very small-energy difference expected between the FM and AF states. **Figure 5a$_1$–a$_4$** shows that TM 3d levels dominate the density of states (DOS) at the Fermi energy and overlap with O 2p states. This indicates that there is a strong interaction between TM and the neighboring O atoms, which results into opposite magnetic moments of O atoms. The contribution to the moment coming from TM 3d orbitals is 2.859μ_B, 3.930μ_B, 3.189μ_B, 2.095μ_B, and 1.015μ_B for TM = Mn, Cr, Fe, Co, and Ni, respectively. In the ground state configuration, the AF state is found to be lower in energy by −0.094, −0.601, −0.832, −0.098, and − 0.102 eV than the FM state for $Zn_{0.929}TM_{0.071}O$ with TM = Cr, Mn, Fe, Co, and Ni, respectively.

3.3.2 Densities of states of Gd ions in ZnO

Figure 5b shows the first-principle calculation for $Zn_{46}O_{48}Gd_2$ and $Zn_{46}O_{47}Gd_2$ nanowires with and without oxygen vacancies, V_O [39]. The spin-up and spin-down DOSs of doped nanowire are significantly different from that of the pristine nanowire. The majority Gd f states (spin-up) that are located well below the valence band maximum are fully occupied. However, the minority unoccupied Gd f states are localized in the vicinity of the Fermi level. The Gd d states in the conduction band overlap with the Gd f states. The hybridization of Gd f states with the states of

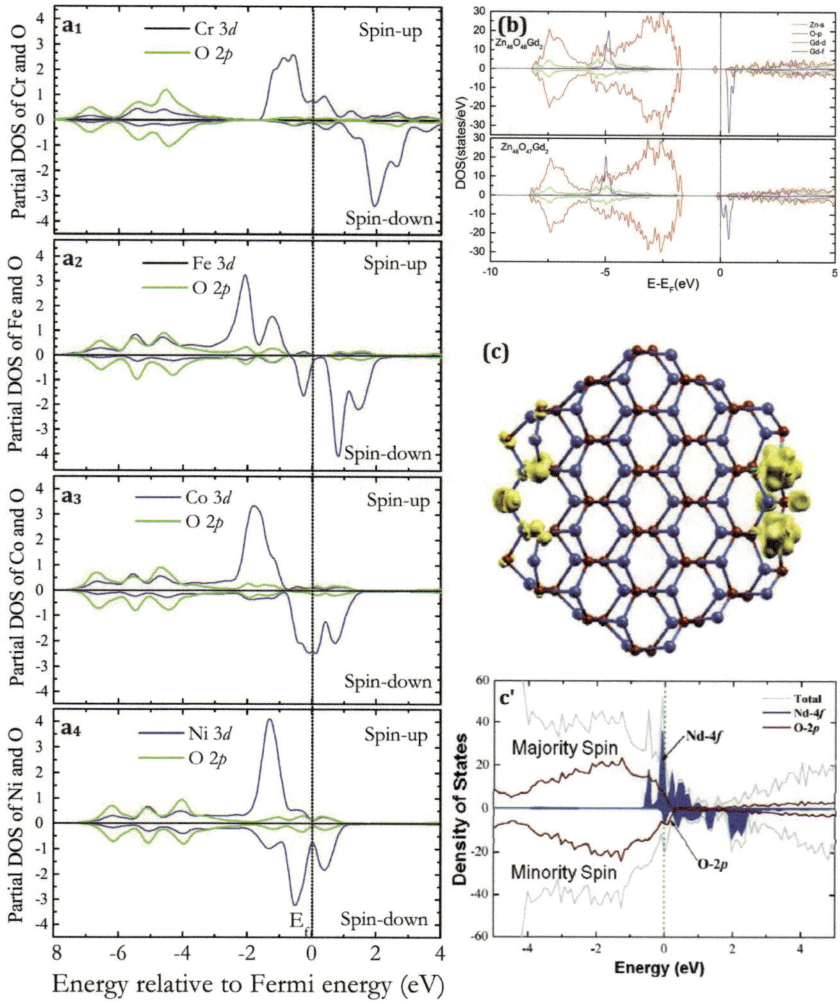

Figure 5.
(a_1–a_4) partial spin DOS of TM 3d and O 2p in $Zn_{24}TM_4O_{28}$ supercell. (b) DOS of $Zn_{46}O_{48}Gd_2$ and $Zn_{46}O_{47}Gd_2$ supercell. (c, c′) Total and projected DOS of $Zn_{51}V_{Zn}O_{54}Nd_2$ nanowire. Fermi level spin-density isosurface (green dotted), Zn(blue), O(red), and Nd(green spheres) (adapted from [19, 28, 39]).

the host material occurs at the Fermi level. By introducing oxygen vacancies, the Gd f state near the Fermi energy becomes partially occupied by donor electrons. Consequently, the carrier concentration around V_O is increased, which mediates the interaction between the s (mostly from Zn) and f electrons. This is evident from the DOS, as the s-f coupling is more prominent than p-f and f-f couplings. For such case, the carrier involved long-range ferromagnetic order to determine the exchange interactions in DMS ZnO. Moreover for these Gd-doped ZnO, oxygen vacancies donate two electrons to the system, mediating the ferromagnetic exchange, and hence, the s-f coupling is more prominent than other mechanisms.

3.3.3 Giant anisotropy in Nd/ZnO nanowire

In **Figure 5c**, the origin of the giant magnetic moment and anisotropy at atomic level is performed by spin-polarized DFT calculations on $Zn_{52}O_{54}Nd_2$ nanowire model using spin density ($\Delta\rho = \rho\uparrow - \rho\downarrow$) and the projection of DOS onto the O-2p and Nd-4f orbitals [28]. The total magnetic moment is high as 6 μ_B calculated from supercell and 3 μ_B per unit cell, and the two Nd atoms are ferromagnetically coupled. It is found that the magnetism mainly comes from the 4f electrons of Nd ions with the local spin moment of ~3 μ_B, and both Zn and O atoms have nearly zero spin contribution. Moreover, significant hybridization is observed between Nd 4f and O 2p orbitals, which leads to the superexchange interaction between two magnetic Nd ions mediated by the nonmagnetic O ions. Both O and Zn vacancies are considered, and it is found that V_{Zn} can enhance the magnetism of about 1 μ_B as compared with defect-free system. This enhanced magnetism mainly comes from the unsaturated 2p orbitals of the surrounding O atoms.

3.4 DMS ZnO with TM = Cr and Mn ions

3.4.1 RTFM in $Zn_{0.94}Cr_{0.06}O$ nanorods

The $Zn_{0.94}Cr_{0.06}O$ nanorods were synthesized by a radio frequency magnetron sputtering deposition technique at different substrate temperatures [46]. The Cr K-edge X-ray absorption near-edge structure and X-ray photoelectron spectroscopy (XPS) results revealed that the Cr^{3+} ions are located at the substitutional Zn sites. The magnetization versus the magnetic field (M-H) loops of $Zn_{0.94}Cr_{0.06}O$ nanorods measured at room temperature is shown in **Figure 6a**. The moment per Cr atom increases with the increasing substrate temperature. The sample prepared at room temperature has a net moment of 0.76 μ_B/Cr. With increasing substrate temperature to 650°C, the value of magnetic moment shows a remarkable increase to 1.16 μ_B/Cr. The coercive field, H_c, of $Zn_{0.94}Cr_{0.06}O$ nanorod grown at room temperatures, 300, 500, and 650°C, are around 104, 42, 53, and 82 Oe, respectively.

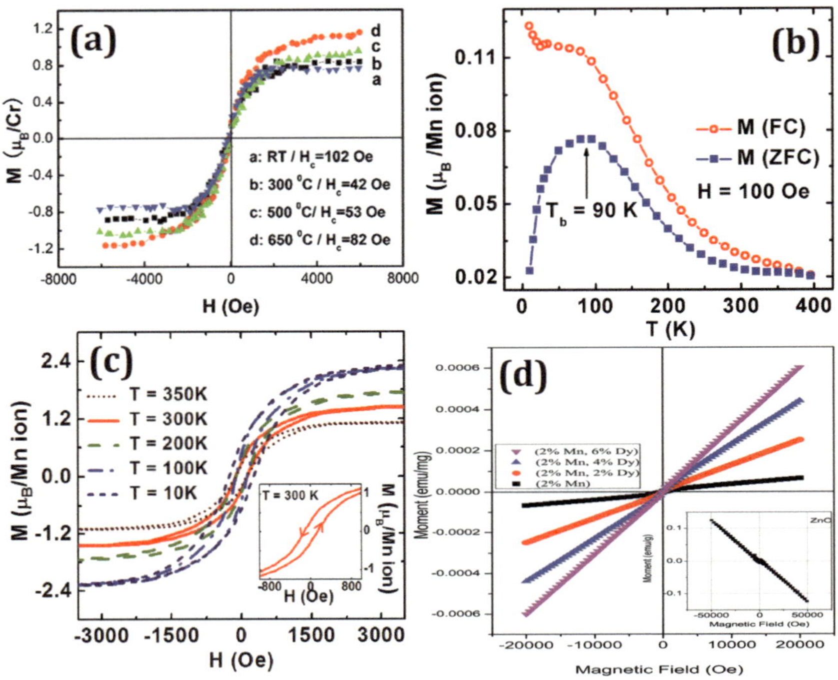

Figure 6.
(a) M-H hysteresis of $Zn_{0.94}Cr_{0.06}O$ nanorods. SQUID measurements: Magnetic moment (b) with temperature (c) with applied field, for ZnO/Mn nanowires. (d) M-H plots for ZnO/(Mn, Dy) nanoparticles (adapted from [24, 44, 46]).

3.4.2 Temperature-dependent magnetization in Mn(1 atom%)/ZnO nanowires

The Mn(1 atom%)-doped ZnO nanowires were synthesized by a gas phase surface diffusion process using MBE system [44]. **Figure 6c** shows the M-H hysteresis loops measured at T = 10, 100, 200, 300, and 350 K for an assembly of Mn-doped ZnO nanowires. The extracted M_s is 2.2 μ_B/Mn ion at 10 K and reduces to 1.4 μ_B/Mn ion at 300 K. Both values are smaller than the theoretical value of 5 μ_B/Mn ion of Mn^{2+} state [1]. The temperature-dependent magnetization (**Figure 6b**) via ZFC and FC at H = 100 Oe shows a typical FM behavior while no intersection is observed in the temperature region of 10–400 K, which reaffirms that T_c is higher than 400 K. However, these FC/ZFC curves show the blocking temperature at T_b = 90 K. The existence of the blocking temperature may result from intrinsic defects, such as oxygen vacancies [47], which contribute weak intrinsic ferromagnetism. The bifurcation begins to increase as the temperature goes below 100 K, and the effect of the external magnetic field starts to overcome the thermal fluctuation and dominate the overall magnetization when the temperature is lower than 100 K.

3.4.3 Magnetism with simultaneous doping from Mn and Dy in DMS ZnO

Figure 6d shows the magnetic results at room temperature with simultaneous doping of Mn and Dy in ZnO nanoparticles prepared by sol–gel process (Mn = 0 and 2% and Dy = 0, 2, 4, and 6%) [24]. The M-H results show that as doping concentration of Dy is increased, magnetic behavior changes from weak ferromagnetic/superparamagnetic to ferromagnetic states. The observed magnetic behavior is linked with oxygen vacancies as determined with EXAFS and PL measurements. The oxygen vacancy-mediated exchange interaction between the Dy^{3+} ions is due to the formation of BMPs.

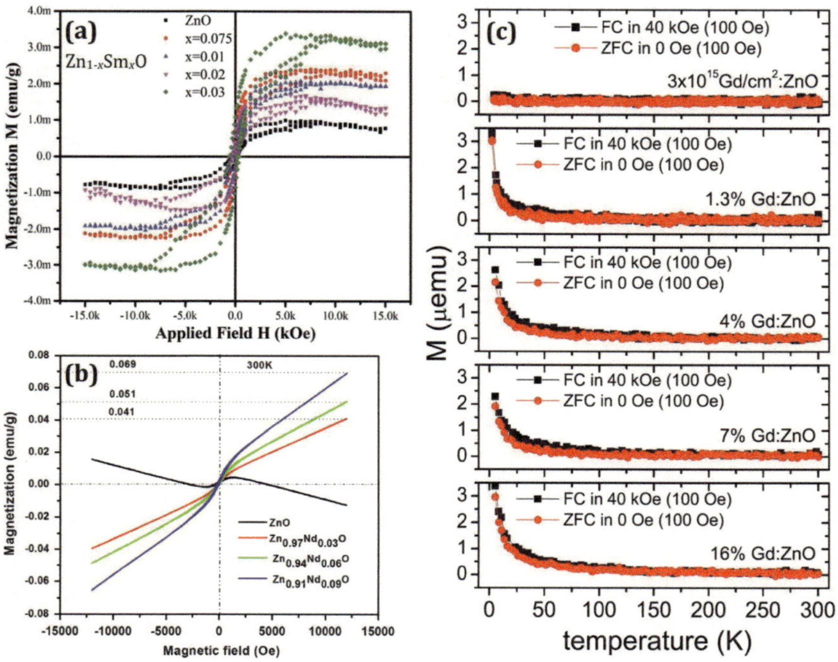

Figure 7.
(a) M-H hysteresis for ZnO/Sm nanoparticles. (b) M-H hysteresis for ZnO/Nd. (c) SQUID M(T) behavior for ZnO/Gd (adopted from [29, 42, 43]).

3.5 DMS ZnO with RE ions

3.5.1 RTFM in Sm/ZnO

The RTFM is enhanced with Sm doping into ZnO is given by M-H hysteresis at room temperature (**Figure 7a**) [29]. It infers that ferromagnetism is intrinsic and formed due to the percolation of BMPs. These BMPs are made up with magnetic cations and defect carrier. A very weak ferromagnetism is observed in pristine ZnO, which is the effect of Zn_i and/or oxygen vacancy defects rather Zn vacancies (V_{Zn}). Because the formation energy of V_{Zn} is too high, it is not preferably formed in ZnO [48].

3.5.2 RTFM in Nd/ZnO

The pure and Nd-doped ZnO nanoparticles were synthesized by the coprecipitation method, and the magnetic results are shown by M-H hysteresis (**Figure 7b**) [42]. All the M-H hysteresis exhibited weak ferromagnetism at room temperature. However, the magnetization increases with increasing Nd^{3+} concentration. The value of saturation magnetization, M_S, is (emu g^{-1}) = 0.041, 0.051, and 0.069, respectively, for $Zn_{0.97}Nd_{0.03}O$, $Zn_{0.94}Nd_{0.06}O$, and $Zn_{0.91}Nd_{0.09}O$. The concentration of oxygen vacancies has a major role in mediating FM exchange interaction among Nd^{3+} ions. It is revealed that O vacancies and Zn interstitials are generated with an increase in Nd^{3+} doping to induce long-range ferromagnetism consistent with the BMP model. Moreover, the s-f coupling between the RE ions (f) and the ZnO host(s) states contributed ferromagnetism of DMSs [14]. The coercivity is also increased with Nd^{3+} concentration.

3.5.3 Temperature-dependent magnetization in Gd/ZnO

Ney et al. [43] reported that for small doping concentrations (1.3% Gd), a large fraction of the Gd atoms is substitutional on Zn lattice sites within wurtzite structure. The magnetic behavior is purely paramagnetic with magnetic moment 7 μ_B/Gd. **Figure 7c** shows the temperature-dependent magnetization from SQUID measurement for Gd-doped ZnO with different Gd concentrations using FC and ZFC conditions [43]. No separation between FC and ZFC magnetization occurs at any temperature, which provides no evidence for ferromagnetic-like behavior. Therefore, all samples have to be considered as paramagnetic. This is in contrast to previous work, where signs of ferromagnetic-like behavior are found for Gd-doped ZnO [49].

3.6 DMS ZnO with Fe and La ions

3.6.1 RTFM in Fe/ZnO nanorods

The $Zn_{1-x}Fe_xO$ (ZFO) [x = 0.01 (ZFO1), 0.03 (ZFO3), and 0.05 (ZFO5)] nanorods were synthesized by a sol-gel process [50]. The XRD pattern revealed the hexagonal wurtzite structure with Fe doping. TEM images show nanorod formation with an average diameter, D(nm) = 10, 48, 14, and 12, and length, L(nm) = 23, 113, 50, and 30, respectively, for ZnO, ZFO1, ZFO3, and ZFO5. **Figure 8a** shows M-H hysteresis for pure and Fe-doped ZnO at room temperature. Pure ZnO exhibits diamagnetic behavior, whereas ZFO samples display superferromagnetic behavior. The values of are M_s(emu g^{-1}) = 0.233, 0.459, and 0.328, respectively, measured for ZFO1, ZFO3, and ZFO5 nanorods. The variations in M_S values depend on factors

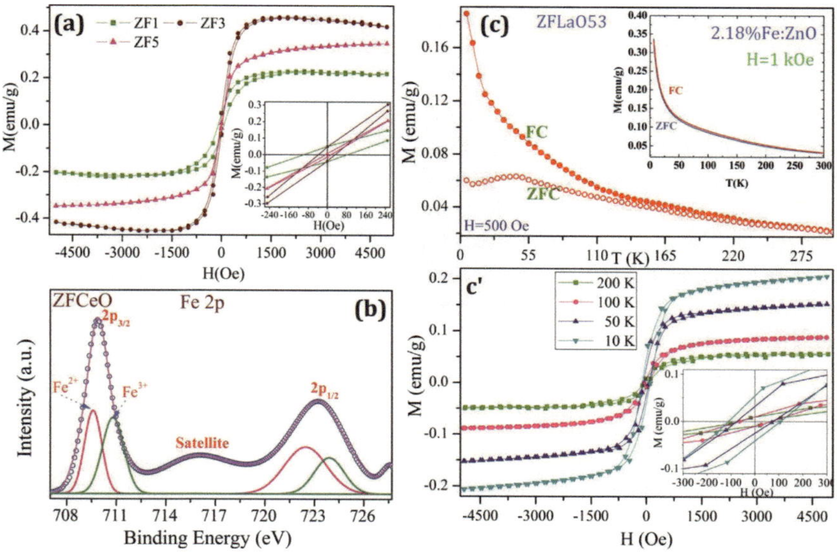

Figure 8.
(a) M-H hysteresis at room temperature for $Zn_{1-x}Fe_xO$ nanorods. (b) XPS spectrum of $Zn_{0.94}Fe_{0.03}Ce_{0.03}O$ nanostructures. (c) M(T) and (c′) M(H) for $Zn_{0.92}Fe_{0.05}La_{0.03}O$ nanoparticles (adapted from [21, 40, 50]).

like shape and size of nanostructures, concentration of dopants, and lattice defects. Furthermore, an F-center exchange (FCE) mechanism has been employed to illustrate the ferromagnetism of Fe-doped ZnO nanorods [51]. In this mechanism, the $Fe^{3+} - V_O^{2-} - Fe^{3+}$ group is common for which an electron is trapped in the oxygen vacancy to make an F-center, where the electron occupies an orbital (p_z), which overlaps the d_z^2 orbital of the d shells of both iron neighbors. The interactions between the ferromagnetic and paramagnetic or antiferromagnetic components lead to variations in saturation magnetization. Another mechanism is related with BMP formation by the alignment of the spins in TM ions [14].

3.6.2 Valence states of Fe in DMS ZnO influenced magnetic ordering

Figure 8b shows Fe 2p XPS spectra in a binding energy 707–728 eV of $Zn_{0.94}Fe_{0.03}Ce_{0.03}O$ (ZFCeO) nanoparticles to find their contribution into ferromagnetism [40]. The Fe^{2+} and Fe^{3+} $2p_{3/2}$ peaks always show satellite peaks at 6 and 8 eV above the principal peaks at 709.5 and 711.2 eV, respectively. The satellite peak is found in energy region of 6–8 eV above $2p_{3/2}$ principal peak, which indicates that ZFCeO DMS has Fe coexisting in both Fe^{2+} and Fe^{3+} states. For this, a multiple fitting of Fe 2p peaks with satellites show peaks corresponding to Fe^{2+} (709.60 and 722.51 eV) and Fe^{3+} (710.82 and 723.97 eV). It indicates that the Fe ions have mixed valences of +2 and + 3. The peaks related with $2p_{3/2}$ 709.89 eV and $2p_{1/2}$ 723.35 eV are also observed. Therefore, it is found that the Fe exists in mixed Fe^{2+} and Fe^{3+} oxidation states to give RTFM due to Fe^{2+}-Fe^{3+} transitions via oxygen vacancies.

3.6.3 Magnetic ordering with La ions in Fe/ZnO nanoparticles

The magnetic results for $Zn_{0.95}Fe_{0.05}O$ (ZFO5) and $Zn_{0.92}Fe_{0.05}La_{0.03}O$ (ZFLaO53) with M_s(emu g^{-1}) = 0.328 and 0.044 and M_r(emu g^{-1}) = 0.0083 and 0.0064 with H_c(Oe) = 12 and 144, respectively, are reported. The origin of observed magnetism at room temperature for La-doped ZFO5 is described via ZFC and FC magnetization SQUID measurement. **Figure 8c** shows temperature-dependent ZFC and FC measurement with H = 500 Oe. The superimposition of ZFC/FC plots between 150 and 300 K, as well as their clear separation at low temperature with blocking temperature, T_B is observed. The observed T_B might correspond with Néel temperature, T_N (~42 K) of AF [52]. For more detail, M-H hysteresis is also measured at 200, 100, 50, and 10 K (**Figure 8c′**). The values M_s and M_r are enhanced with temperature when going from 300 to 10 K. This is due to the exchange interaction from AF to FM states. It is also shown that for 200–50 K, H_c varies so slowly, but at 10 K, it abruptly increased to 117 Oe, which is smaller than 144 Oe that is observed at room temperature. It means after AF transition, there is some possibility of FM clustered growth in ZFLaO sample [53]. The localization of electrons in magnetic clusters leads to develop high-spin and low-spin intersite electronic transitions. These magnetic clusters may also result from magnetic polarons [54].

3.7 DMS ZnO with Co, La, Gd, and Ce ions

3.7.1 RTFM in La- and Gd-doped $Zn_{0.95}Co_{0.05}O$ nanostructure

Figure 9a shows the M-H hysteresis for $Zn_{0.95}Co_{0.05}O$ (ZCO5), $Zn_{0.92}Co_{0.05}La_{0.03}O$ (ZCLO53), and $Zn_{0.92}Co_{0.05}Gd_{0.03}O$ (ZCGO53) nanostructure, measured at room temperature [36]. The pure ZCO5 shows weak ferromagnetism of M_s(emu g^{-1}) = 0.354 and M_r(emu g^{-1}) = 0.0276 with H_c(Oe) = 40 Oe. However, the La- and Gd-doped ZCO5 result into paramagnetic-type behavior. The weak ferromagnetism in ZCO5 exists due to antiferromagnetic, AF interactions among Co^{2+} ions [41, 55]. The AF coupling between Co impurities is favored when Co atoms are separated by more than a ZnO unit. While the ferromagnetic coupling is stable if AF interaction in neighboring Co–Co ions falling into contour of BMPs. However, the observed paramagnetism in La- and Gd-doped ZCO5 is related with

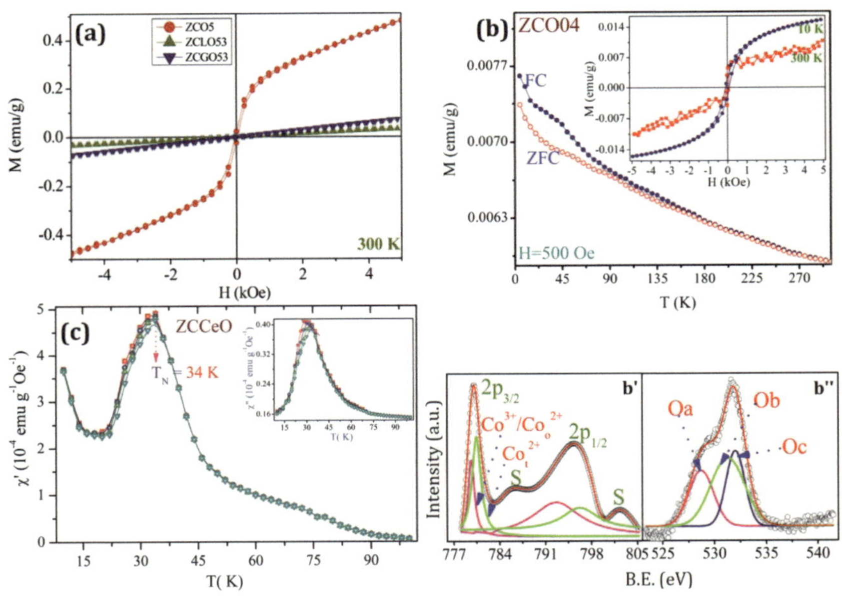

Figure 9.
(a) M-H hysteresis for Co-, La-, and Gd-doped ZnO nanoparticles, measured at room temperature. (b) M(T) and M(H) (inset) for $Zn_{0.996}Co_{0.004}O$ (ZCO04) nanoparticles. (b′ and b″) XPS spectra for Co 2p and O 1 s. (c) Temperature-dependent AC magnetic susceptibility (χ) of ZFCeO nanoparticles (adopted from [15, 36, 40]).

AF interaction establishment following 4f-5d-3d transition in Co and RE ions, suggested by Singh et al. [56].

3.7.2 Magnetic ordering in $Zn_{0.996}Co_{0.004}O$ nanoparticles

The $Zn_{0.996}Co_{0.004}O$ (ZCO04) nanoparticles synthesized with sol–gel process for which free-charge carriers and oxygen vacancies might induce long-range ferromagnetic ordering [15]. The XRD pattern results into wurtzite structure of ZCO04. The ZCO04 crystalline product has nanorod formation with D(nm) = 23 ± 3 and L(nm) = 57 ± 5. **Figure 9b** (inset) showed the RTFM of M_s (emu g^{-1}) 0.0062 and M_r (emu g^{-1}) = 0.0038 with H_c = 54 Oe. However, the pure ZnO nanorods are diamagnetic [50]. Xu et al. [57] reported RTFM with higher surface-to-volume ratio of nanostructure, which contribute large amount of surface oxygen vacancies defects. It is expected that the RTFM is attributed via exchange interactions among unpaired electron spins arising from either vacancies or surface defects, which is explained on the basis of donor impurity band exchange model form BMPs [58]. It is theoretically investigated that the oxygen vacancies have remarkable change in band structure of host oxides to induce ferromagnetism [59]. For this case of BMPs, the electrons are locally trapped by oxygen vacancies, with the trapped electron occupying an orbital overlapping with the *d* shells of Co neighbors.

To evaluate the origin of RTFM of ZCO04 nanoparticles, the temperature-dependent magnetization is given in **Figure 9b** via ZFC and FC at H = 500 Oe. The separation between ZFC and FC starts increasing with reducing temperature from 300 to 5 K which indicates antiferromagnetic interactions converted to ferromagnetic state. The absence of blocking temperature in ZFC might indicate long-range antiferromagnetism without any cluster growth. The exchange interactions between neighboring magnetic ions mediated by an F-center form a BMP contributing long-range ferromagnetism. At 10 K, the magnetic hysteresis is also shown in the inset of **Figure 9b** with M_s(emu g^{-1}) = 0.0154 and M_r(emu g^{-1}) = 0.002 with H_c(Oe) = 93.

3.7.3 Valence states of Co and O ions in $Zn_{0.996}Co_{0.004}O$ nanoparticles

Figures 9b′, b″ shows the XPS spectra for Co 2p and O 1 s of $Zn_{0.996}Co_{0.004}O$ (ZCO04) nanoparticles. For Co 2p, the doublet is the spin-orbit coupling ($2p_{3/2}$ and $2p_{1/2}$) given in **Figure 9b′**. The values of binding energy, Co $2p_{3/2}$ ~ 780.019 eV, $2p_{1/2}$ ~ 795.51 eV, and ΔE ~ 15.51 eV, and satellite peak (S) ~ 785.48 eV are observed. The binding energies of Co $2p_{3/2}$ and $2p_{1/2}$ indicate that the Co ions exist either in +3 or + 2 valence states [60]. The difference ΔE of binding energy among Co $2p_{3/2}$ and $2p_{1/2}$ levels corresponds well with Co^{2+} that is homogeneously surrounded by oxygen in tetrahedral coordination [61]. However, the peak S is found in the energy region of 6–8 eV above the principle peak Co $2p_{3/2}$ and the value of S ~ 6 eV. It indicates the formation of multiple coordinations, i.e., tetrahedral or octahedral Co^{2+} ions. For more clarification, Co 2p peaks shown by octahedral Co^{2+} (Co_o^{2+}) and Co^{3+} and tetrahedral Co^{2+} (Co_t^{2+}) are clearly marked.

To find defects/vacancies in ZCO04, the O 1 s spectra is shown in **Figure 9b″**, which deconvoluted into three peaks (O_a, O_b, O_c) [15]. The peak located on low-binding energy side, O_a ~ 528.79 eV, is attributed with O^{2-} ions in wurtzite structure. This O_a of O 1 s is associated with Zn-O bonds. The O_b peak at 531.19 eV is associated with O^{2-} ions in the oxygen-deficient regions within the ZnO matrix, which indicate defect formation. The O_c peak at 532.16 eV is attributed to chemisorbed oxygen on the surface of the ZnO.

3.7.4 AC magnetic susceptibility (χ) of $Zn_{0.94}Fe_{0.03}Ce_{0.03}O$ nanoparticles

The temperature-dependent real, $\chi'(T)$, and imaginary, $\chi''(T)$, components of the AC magnetic susceptibility of Ce-doped ZFO nanoparticles at frequencies 100, 300, 500, and 1000 Hz are shown in **Figure 9c** [40]. The frequency-independent peak maxima of χ' correspond to magnetic phase transition into an AF state with T_N = 56 K is observed. A slight dispersion in the $\chi'(T)$ after the peak maxima is also observed. It indicates the existence of certain spin glass, ferromagnetic clusters and cluster glass magnetic type states [62]. However, the peak of $\chi''(T)$ increases with decreasing frequency. This is qualitatively different from the behavior of most spin glasses in which we expect an increase of the peak magnitude with increasing frequency. Thus, the obtained AC magnetization data supports the formation of the AF DMS materials with a certain cluster glass formation.

4. Conclusion

The spintronic materials are the ferromagnetic oxide semiconductors for which both charge and spin degrees of freedom exist, but with weak ferromagnetism and small coercivity, there is a limitation in practical applications. Thus, these DMSs are widely characterized for the development of semiconductor devices, which create a new dimensionality to control and achieve high T_C ferromagnetism. The first principle calculation revealed that the Cr, Fe, Co, and Ni dopants in ZnO prefer to occupy surface sites instead of bulk sites, while Mn exhibits no site preference and distributes uniformly in ZnO. In addition, the TM ions in ZnO are intrinsically AF because their substitution at Zn sites does not introduce any extra carriers. The interaction among localized spins on the TM ions and delocalized carrier electrons originating from the O vacancies is responsible for the required magnetic transition. The doping RE ions with intrinsic strong magnetic anisotropy and tailoring the coupling between dopants and defects should have a general approach toward a stable ferromagnetic order in DMS ZnO. The observed magnetism of DMS ZnO for both TM and RE ions might result with BMPs in oxygen vacancies.

Acknowledgements

This research work is financially supported by CSIR India, for giving a Senior Research Associate fellowship (Pool Scientist; letter No. B-12287).

Author details

Kuldeep Chand Verma[1,2]

1 Department of Physics, Panjab University, Chandigarh, India

2 CSIR-Central Scientific Instruments Organisation, Chandigarh, India

*Address all correspondence to: dkuldeep.physics@gmail.com

References

[1] Dietl T, Ohno H, Matsukura F, Cibert J, Ferrand D. Zener model description of ferromagnetism in zinc-blende magnetic semiconductors. Science. 2000;**287**: 1019-1022

[2] Furdyna JK. Diluted magnetic semiconductors. Journal of Applied Physics. 1988;**64**:R29-R56

[3] Chanda A, Gupta S, Vasundhara M, Joshi SR, Muttae GR, Singh J. Study of structural, optical and magnetic properties of cobalt doped ZnO nanorods. RSC Advances. 2017;7: 50527-50536

[4] Kittilstved KR, Liu WK, Gamelin DR. Electronic structure origins of polarity-dependent high-T_C ferromagnetism in oxide-diluted magnetic semiconductors. Nature Materials. 2006;**5**:290-297

[5] Sharma P, Gupta A, Rao KV, Owens FJ, Sharma R, Ahuja R, et al. Ferromagnetism above room temperature in bulk and transparent thin films of Mn-doped ZnO. Nature Materials. 2003;**2**:673-677

[6] Ohno H. A window on the future of spintronics. Nat. Mat. 2010;**9**:952-954

[7] Saitoh E. New order for magnetism. Nat. 2008;**455**:474-475

[8] Ohno H, Chiba D, Matsukura F, Omiya T, Abe E, Dietl T, et al. Electric-field control of ferromagnetism. Nature. 2000;**408**:944-946

[9] Chiba D, Yamanouchi M, Matsukura F, Ohno Electrical H. Manipulation of magnetization reversal in a ferromagnetic semiconductor. Science. 2003;**301**:943-945

[10] Dietl T. A ten-year perspective on dilute magnetic semiconductors and oxides. Nature Materials. 2010;**9**: 965-974

[11] Gupta JA, Knobel R, Samarth N, Awschalom DD. Ultrafast manipulation of electron spin coherence. Science. 2001;**292**:2458-2461

[12] Chambers S. Is it really intrinsic ferromagnetism? Nature Materials. 2010;**9**:956-957

[13] Kundaliya DC, Ogale SB, Lofland SE, Dhar S, Metting CJ, Shinde SR, et al. On the origin of high-temperature ferromagnetism in the low-temperature processed Mn-Zn-O system. Nature Materials. 2004;**3**: 709-714

[14] Coey JMD, Venkatesan M, Fitzgerald CB. Donor impurity band exchange in dilute ferromagnetic oxides. Nature Materials. 2005;**4**: 173-179

[15] Verma KC, Bhatia R, Kumar S, Kotnala RK. Vacancies driven magnetic ordering in ZnO nanoparticles due to low concentrated Co ions. Mater. Res. Exp. 2016;**3**:076103-076112

[16] Zhen WZ, Min ZJ, Gao HZ, Xiong LW. Effect of oxygen vacancy defect on the magnetic properties of Co-doped ZnO. Chinese Physics B. 2011;**20**: 027103

[17] Kumar V, Ntwaeaborwa OM, Soga T, Dutta V, Swart HC. Rare earth doped zinc oxide nanophosphor powder: A future material for solid state lighting and solar cells. ACS Photonics. 2017;**4**:2613-2637

[18] Venkatesan M, Fitzgerald CB, Lunney JG, Coey JMD. Anisotropic ferromagnetism in substituted zinc oxide. Physical Review Letters. 2004;**93**: 177206-14

[19] Wang Q, Sun Q, Jena P, Kawazoe Y. Magnetic properties of transition-metal-doped $Zn_{1-x}T_xO$ (T = Cr, Mn, Fe, Co,

and Ni) thin films with and without intrinsic defects: A density functional study. Physical Review B. 2009;**79**: 115407-115413

[20] Park YS, Litton CW, Collins TC, Reynolds DC. Exciton Spectrum of ZnO. Physics Review. 1966;**143**:512-519

[21] Verma KC, Kotnala RK. Oxygen vacancy induced by La and Fe into ZnO nanoparticles to modify ferromagnetic ordering. Journal of Solid State Chemistry. 2016;**237**:211-218

[22] Gao D, Zhang Z, Fu J, Xu Y, Qi J, Xue D. Room temperature ferromagnetism of pure ZnO nanoparticles. Journal of Applied Physics. 2009;**105**:113928-4

[23] Srinivasulu T, Saritha K, Ramakrishna Reddy KT. Synthesis and characterization of Fe-doped ZnO thin films deposited by chemical spray pyrolysis. Modern Elect. Mater. 2017;**3**: 76-85

[24] Tiwari N, Kumar S, Ghosh AK, Chatterjee S, Jhaa SN, Bhattacharyya D. Structural investigations of (Mn, Dy) co-doped ZnO nanocrystals using X-ray absorption studies. RSC Advances. 2017; 7:56662-56675

[25] Valerio LR, Mamani NC, de Zevallos AO, Mesquita A, Bernardi MIB, Doriguetto AC, et al. Preparation and structural-optical characterization of dip-coated nanostructured Co-doped ZnO dilute magnetic oxide thin films. RSC Advances. 2017;7:20611

[26] Samanta A, Goswami MN, Mahapatra PK. Magnetic and electric properties of Ni-doped ZnO nanoparticles exhibit diluted magnetic semiconductor in nature. Journal of Alloys and Compounds. 2018;**730**: 399-407

[27] Verma KC, Kotnala RK. Understanding lattice defects to influence ferromagnetic order of ZnO nanoparticles by Ni, Cu, Ce ions. Journal of Solid State Chemistry. 2017;**246**: 150-159

[28] Wang D, Chen Q, Xing G, Yi J, Bakaul SR, Ding J, et al. Robust room-temperature ferromagnetism with Giant anisotropy in Nd-doped ZnO nanowire arrays. Nano Letters. 2012;**12**: 3994-4000

[29] Arora D, Asokan K, Mahajan A, Kaura H, Singh DP. Structural, optical and magnetic properties of Sm doped ZnO at dilute concentrations. RSC Advances. 2016;**6**:78122-78131

[30] Li H, Huang Y, Zhang Q, Qiao Y, Gu Y, Liu J. Facile synthesis of highly uniform Mn/Cocodoped ZnO nanowires: Optical, electrical, and magnetic properties. Nanoscale. 2011;**3**: 654-660

[31] Hou L, Zhou W, Zou B, Zhang Y, Han J, Yang X. Spin exciton interaction and related microphotoluminescence spectra of ZnSe:Mn DMS nanoribbon. Nanotechnology. 2017;**28**:105202-105211

[32] Midya N, Neogi SK, Ahmed Md A, Banerjee A, Kumar P, Kanjilald D. Correlation between magnetic and microstructural properties of low energy ion irradiated and un-irradiated $Zn_{0.95}Mn_{0.05}O$ films. RSC Advances. 2017;7:771-781

[33] Rouchdi M, Salmani E, Fares B, Hassanain N, Mzerd A. Synthesis and characteristics of Mg doped ZnO thin films: Experimental and ab-initio study. Results Phys. 2017;7:620-627

[34] Kaur J, Negi NS, Kotnala RK, Verma KC. Hydrothermal conditions on $Sn_{0.95}Co_{0.05}O_2$: Nanostructures, ferromagnetism and optical behavior. Journal of Sol-Gel Science and Technology. 2013;**65**:411-419

[35] Anjaneyulu C, Naresh G, Kumar VV, Tardio J, Rao TV, Venugopal A. Influence of rare earth

(La, Pr, Nd, Gd, and Sm) metals on the methane decomposition activity of Ni-Al catalysts. ACS Sustainable Chemistry & Engineering. 2015;**3**: 1298-1305

[36] Verma KC, Kotnala RK. Defects due to lattice distortion and nano-size intermediate ferromagnetism in La, Gd substituted. $Zn_{0.95}Co_{0.05}O$. Current Applied Physics. 2016;**16**: 175-182

[37] Sun LW, Shi HQ, Li WN, Xiao HM, Fu SY, Cao XZ, et al. Lanthanum-doped ZnO quantum dots with greatly enhanced fluorescent quantum yield. Journal of Materials Chemistry. 2012;**22**: 8221-8227

[38] Bantounas I, Singaravelu V, Roqan IS, Schwingenschlogl U. Structural and magnetic properties of Gd-doped ZnO. Journal of Materials Chemistry C. 2014;**2**:10331-10336

[39] Aravindh SA, Schwingenschloegl U, Roqan IS. Ferromagnetism in Gd doped ZnO nanowires: A first principles study. Journal of Applied Physics. 2014;**116**: 233906-5

[40] Verma KC, Kotnala RK. Defects-assisted ferromagnetism due to bound magnetic polarons in Ce into Fe, Co: ZnO nanoparticles and first-principle calculations. Physical Chemistry Chemical Physics. 2016;**18**:5647-5657

[41] Inamdar DY, Pathak AK, Dubenko I, Ali N, Mahamuni S. Room temperature ferromagnetism and photoluminescence of Fe doped ZnO Nanocrystals. Journal of Physical Chemistry C. 2011;**115**:23671-23676

[42] Vijayaprasath G, Murugan R, Mahalingam T, Hayakawa Y, Ravi G. Enhancement of ferromagnetic property in rare earth neodymium doped ZnO nanoparticles. Cer. Int. 2015;**41**: 10607-10615

[43] Ney V, Ye S, Kammermeier T, Ollefs K, Wilhelm F, Rogalev A, et al. Structural and magnetic analysis of epitaxial films of Gd-doped ZnO. Physical Review B. 2012;**85**: 235203-235207

[44] Chang LT, Wang CY, Tang J, Nie T, Jiang W, Chu CP, et al. Electric-field control of ferromagnetism in Mn-doped ZnO nanowires. Nano Letters. 2014;**14**: 1823-1829

[45] Ram M, Verma KC, Negi NS. Room temperature ferromagnetism and photoluminescence of nanocrystalline $Zn_{1-x}Co_xO$ thin films prepared by sol-gel MOD. Phil. Mag. Lett. 2014;**94**(7): 404-414

[46] Jin CG, Yang Y, Wu ZF, Zhuge LJ, Han Q, Wu XM, et al. Tunable ferromagnetic behavior in Cr doped ZnO nanorod arrays through defect engineering. Journal of Materials Chemistry C. 2014;**2**:2992-2997

[47] Wang Q, Sun Q, Chen G, Kawazoe Y, Jena P. Vacancy-induced magnetism in ZnO thin films and nanowires. Physical Review B. 2008;**77**: 205411

[48] Phan TL, Zhang YD, Yang DS, Nghia NX, Thanh TD, Yu SC. Defect-induced ferromagnetism in ZnO nanoparticles prepared by mechanical milling. Applied Physics Letters. 2013; **102**(7):072408

[49] Ney V, Ye S, Kammermeier T, Ney A, Zhou H, Fallert J, et al. Structural, magnetic, and optical properties of Co- and Gd-implanted ZnO(0001) substrates. Journal of Applied Physics. 2008;**104**:083904

[50] Kaur J, Kotnala RK, Gupta V, Verma KC. Anionic polymerization in Co and Fe doped ZnO: Nanorods, magnetism and photoactivity. Current Applied Physics. 2014;**14**:749-756

[51] Verma KC, Singh J, Ram M, Sharma DK, Sharma A, Kotnala RK. Enhancement in the magnetic, optical and electrical properties of $Ti_{0.97}Co_{0.03}O_2$ and $Ti_{0.97}Fe_{0.03}O_2$ nanoparticles with Ce co-doping. Physica Scripta. 2012;**86**:025704

[52] Fan JP, Li XL, Quan ZY, Xu XH. Tunable magnetic and transport properties of p-type ZnMnO films with n-type Ga, Cr, and Fe codopants. Applied Physics Letters. 2013;**102**: 102407-3

[53] Gnatchenko SL, Ratner AM, Baran M, Szymczak R, Szymczak H. Formation and interaction of ferromagnetic clusters in antiferromagnetic $YBa_2Cu_3O_{6+x}$ films. Physical Review B. 1997;**55**:3876-3885

[54] Kovaleva NN, Kugel KI, Bazhenov AV, Fursova TN, Loser W, Xu Y, et al. Formation of metallic magnetic clusters in a Kondo-lattice metal: Evidence from an optical study. Scientific Reports. 2012;**2**:890-897

[55] Walsh A, Da Silva JLF, Wei SH. Theoretical description of carrier mediated magnetism in cobalt doped ZnO. Physical Review Letters. 2008; **100**:256401-256404

[56] Singh SK, Rajaraman G. Decisive interactions that determine ferro/ antiferromagnetic coupling in {3d–4f} pairs: A case study on dinuclear {V(IV)-Gd(III)} complexes. Dalton Transactions. 2013;**42**:3623-3630

[57] Xu X, Xu C, Dai J, Hu J, Li F, Zhang S. Size dependence of defect-induced room temperature ferromagnetism in undoped ZnO nanoparticles. Journal of Physical Chemistry C. 2012;**116**:8813-8818

[58] Banerjee S, Mandal M, Gayathri N, Sardar M, et al. Applied Physics Letters. 2007;**91**:182501

[59] Hsu HS, Huang JCA, Huang YH, Liao YF, Lin MZ, Lee CH, et al. Evidence of oxygen vacancy enhanced room-temperature ferromagnetism in Co-doped ZnO. Applied Physics Letters. 2006;**88**:242507

[60] Lu XF, Wu DJ, Li RZ, Li Q, Ye SH, Tong YX, et al. Hierarchical $NiCo_2O_4$ nanosheets@hollow microrod arrays for high-performance asymmetric supercapacitors. Journal of Materials Chemistry A. 2014;**2**:4706-4713

[61] Lee HJ, Jeong SY, Cho CR, Park CH. Study of diluted magnetic semiconductor: Co-doped ZnO. Applied Physics Letters. 2002;**81**:4020

[62] Rathore SS, Vitta S. Effect of divalent Ba cation substitution with Sr on coupled 'multiglass' state in the magnetoelectric multiferroic compound $Ba_3NbFe_3Si_2O_{14}$. Scientific Reports. 2015;**5**:9751

Chapter 7

Ferromagnetism in SnO_2 Doped with Transition Metals (Fe, Mn and Ni) for Spintronics Application: A Review of Experimental Status

Yared Worku, Dipti Ranjan Sahu and Vijaya Vallabhapurapu Srinivasu

Abstract

Due to their potential application in the field of spintronics, the discovery of various types of oxide-based dilute magnetic semiconductors (ODMS) materials that might work at practical room temperature ferromagnetism (RTFM) has recently attracted great attention. Among ODMS materials, transition metal™ doped tin oxide (SnO_2) compounds are important for the investigation of ferromagnetism due to its special important property such as high chemical stability, high carrier density, n-type behavior and trait long range ferromagnetism. However, the question of understanding the mechanism of ferromagnetism (FM) process is still not fully understood in these materials, due to unable to know exactly whether its FM property arises from the nature of the intrinsic property or secondary phases of the material. According to the results from many literature surveys, the mechanism of magnetic ordering responsible for magnetic exchange interaction in these materials is highly affected by oxygen vacancy, defects, dopant types and concentration, temperature, sample preparation method and so on. In this chapter, we reviewed the mechanism of ferromagnetism observed of Ni, Mn and Fe-doped SnO_2 materials.

Keywords: spintronics, diluted magnetic semiconductors (DMS), room temperature ferromagnetism (RTFM), TM doped tin oxide, oxygen vacancy and defect

1. Introduction

Current semiconductor-based electronics device uses only the electron charging property to perform a particular feature in which the electron spin degree is completely ignored [1]. The spin property of an electron which is associated with an intrinsic angular momentum of the electron provides new effects and new functionalities to electronics materials based on Spintronics principle [2, 3]. Spintronics deals with the role played by the spin of an electron associated with its magnetic moment, as well as the charge degree of electron [4]. Spintronics devices have several important applications compared to non-spin based electronics device, such as consume

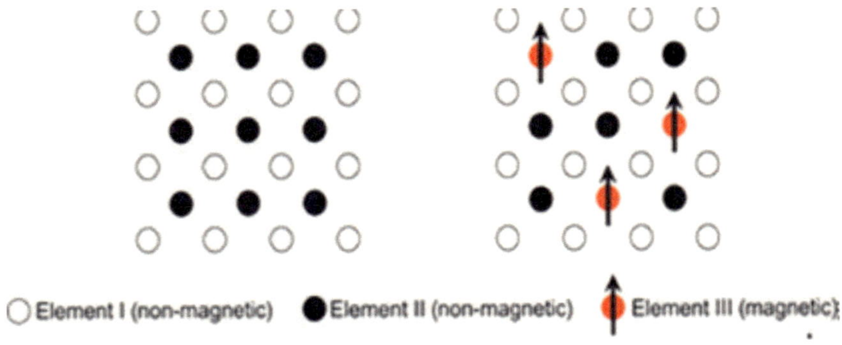

Figure 1.
Schematic view of a non-magnetic (left) semiconductor and a diluted magnetic semiconductor (right) [5].

less power of electricity, fast data processing speed, their memories are non-volatile [4]. Starting from metal-based technology, the research area of Spintronics shifted to the recent development of diluted magnetic semiconductors (DMS) materials which are compatible with standard semiconductor based electronics device. DMS are materials prepared through which a certain amount of the cations in a host semiconductor are partially replaced by transition metal ions (Mn, Ni, Co, Fe, Cr) as shown in **Figure 1** [5] as a result the materials attains both semiconducting and magnetic property which is makes these materials advantageous and applicable for Spintronics application. The total ferromagnetic behavior of these materials is linked to the interaction of the spin of the magnetic ions with the itinerant carriers [6–8]. DMS are important materials in the sense that logic, communications and storage operation can be achieved within the same materials technology [9, 10]. The property of achieving RTFM is one of the most important factors that determined DMS material to be used for practical spintronics application [7], The sp-d exchange mechanism between the d states of the TM doping and sp free carriers as well as the double exchange mechanisms are the main factor in the production of ferromagnetism in ODMS materials between d states of TM ions [11]. Among DMS materials oxide based DMS materials such as TM doped with HfO_2, TiO_2, ZnO and SnO_2 are more advantageous than normal DMS materials and have important magnetic properties arises from a large sp-d exchange interactions between the magnetic ion elements and band electrons [9, 10, 12, 13]. ODMS has important special properties such as having high n-type carrier concentrations wide band gap, light transparency, capability to be grown at low temperatures, ecological safety and cheap [14–16]. Due to its n-type semiconductor, good conductivity, high carrier density and high chemical stability, SnO_2 doped with TM is particularly promising materials for spintronic applications [17, 18]. SnO_2 naturally existing in cassiterite form and it has tetragonal rutile structure and its wide band gap is about 3.6 eV [19, 20]. SnO_2 has many technological applications, including gas sensors, solar cells, heat reflectors, lithium ion batteries and other optoelectronic devices [21–23].

2. Ferromagnetism in oxide-based DMS

In the recent years the research field of RTFM in O-DMS has got more attention and many kinds of compounds have been discovered [24]. However, the idea behind the original source of ferromagnetism in these materials is not well understood a not complete it becomes the most challenging area in solid state physics [25]. Several groups have stated that the mechanism behind ferromagnetism in most O-DMS materials is the material's intrinsic property itself or the direct and indirect interaction between only magnetic impurities and magnetic impurity ions through oxygen vacancies [10, 26–30]. Recently, various experimental methods have been used to study the magnetic properties of DMS materials, in particular the vibrating sample magnetometer (VSM), the superconducting quantum interference device (SQUID), the physical property measurement system (PPMS) and the electron spin resonance (ESR) techniques. According to the results from many literature indicated that sample preparation, growth conditions, dopant type and concentration, co-doping effect, oxygen vacancies, defects and crystal structure has played a role for the magnetic behaviors observed in ODMS material [31–37]. Some scholars reported that vacancy-induced magnetism has been played a major role for the observed ferromagnetism in undoped SnO_2 [38]. In some cases SnO_2 thin films does not shows RTFM when doped with 3*d* cations rather show when doped with Mn, Cr, Fe, Co, or Ni [39–41]. Similarly undoped SnO_2 did not shows FM behavior. However, the doped SnO_2 shows FM behavior at higher doping level completely removes the ferromagnetic behavior of the doped one [31, 42]. As shown in **Figure 2** the improvement of magnetization by co doping (Ni-Mn, Fe-Co, Fe-Ni and Fe-Mn) in tin oxide has been reported and the mechanism of FM is due to double exchange interactions occur via oxygen vacancies [11, 43]. The electronic or lattice defects of the materials associated with the intrinsic nature of the materials can be responsible for the high temperature FM of TM doped SnO_2 [39]. Despite much experimental success, the idea behind FM in most O-DMS is controversial. Here, we present a brief review of the Fe, Ni and Mn Doped SnO_2 system experimental work.

2.1 Ferromagnetism in Mn-doped SnO_2

Mn-doped SnO_2 is an excellent candidate and promising materials for RTFM study, but only very little work has been reported so far compared to others. Among other preparation methods sol-gel preparation technique is best method for preparation of TM doped SnO_2 thin film and nano structures [44, 45].

2.1.1 Experimental

$SnMnO_2$ thin film is prepared by sol-gel method according to the literature reported [44]. The solution was prepared by dissolving a certain amount of tin tetrachloride $SnCl_4$ and manganese nitrate hydrate [Mn $(NO_3)_2 6H_2O$] in distilled water and

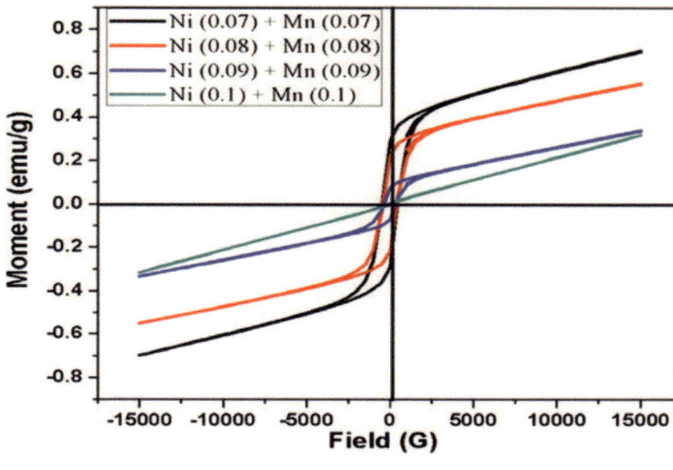

Figure 2.
M-H curves of (Ni-Mn) co-doped with SnO_2 [11].

ethanol respectively and stirring for 5 hour and aging for at least a week, the prepared solution was spin-coated on silicon substrate and heated at 120°C for 25 min. The film precursors were obtained after multilayer coating. Finally, to obtain $SnMnO_2$ thin films, the precursors of films were calcinated in atmospheric air at 5000°C for an hour.

As reported by Tian et al., the chemical co-precipitation method was used to synthesize Mn doped SnO_2 nanoparticles [46]. First, appropriate quantities of $SnCl_2$ and manganese acetic acid were dissolved in ethanol solution, then a few drops of HCl solution were applied to ensure dissolution. Then a 10 M ammonium bicarbonate solution with continuous stirring at 60°C was applied to the mixture solution until a pH of 9 was reached. After being distributed by ultrasonic for 15 min to get nano-crystalline powders. The resulting precipitation washed to clean the impurities and dried in air at 150°C. Finally, the nano crystalline powders were sintered in the air for 3 hours. X-ray diffraction (XRD) recorded the crystal structure of the synthesized $SnMnO_2$ thin film as shown in **Figures 3** and **4**. The study magnetization property and RTFM were performed using a superconductive quantum interference device (SQUID) and vibrating sample magnetometer (VSM) respectively.

2.1.2 Result and discussion

According to various reports, the origin of the observed FM in Mn-doped SnO_2 depends on a number of factors; some reported that Mn- SnO_2 prepared by PLD method exhibits the only paramagnetic behavior [47]. Similarly, others reported that the dopant Mn does not contribute any role for the observed FM behavior of Mn-doped SnO_2 films; it is assumed that oxygen vacancies and defects are the main factors contributing to the FM order in the system as shown in **Figure 5** below [48]. Others report on Mn-SnO_2 powders confirmed that the observed FM property is likely the results from oxygen vacancies, and Mn doping has only a significant role of the observed source of RTFM in the materials [49]. High level of Mn dopant can degrade the FM behavior where as small doping concentration intrinsic defects can act as a source of the FM in Mn-doped SnO_2 due to a very large magnetic moment [39]. On the other hand, a study reported on Mn doped SnO_2 nanoparticles shows sintering temperature and doping concentration can affects the magnetism of the materials system [46]. As shown in **Figure 7** that Mn-doped SnO_2 powder that concentration of Mn ion contributes to a decrease in the average magnetic moment of magnetic ions, this is due to the competition between the super-exchange antiferromagnetic coupling and the F-center coupling mechanism [50]. Similarly, other research on Mn-doped SnO_2 thin films

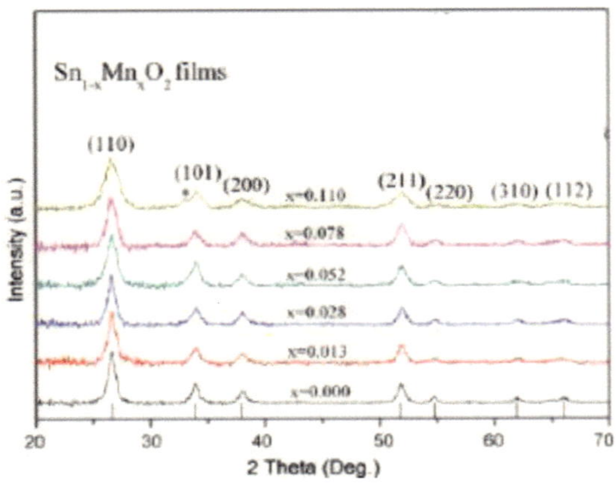

Figure 3.
XRD patterns of $SnMnO_2$ films [44].

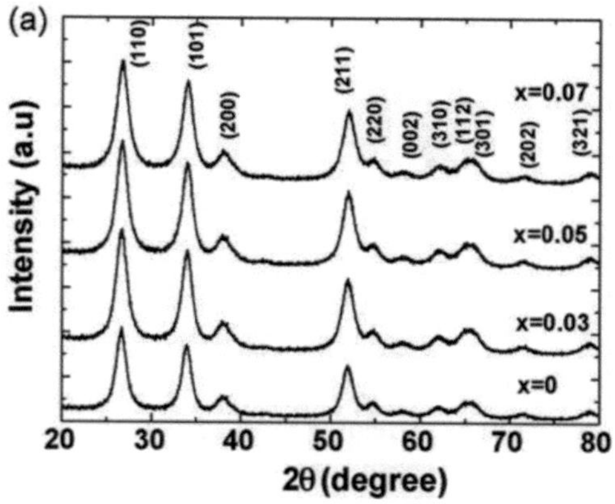

Figure 4.
XRD patterns of $SnMnO_2$ nano particle [46].

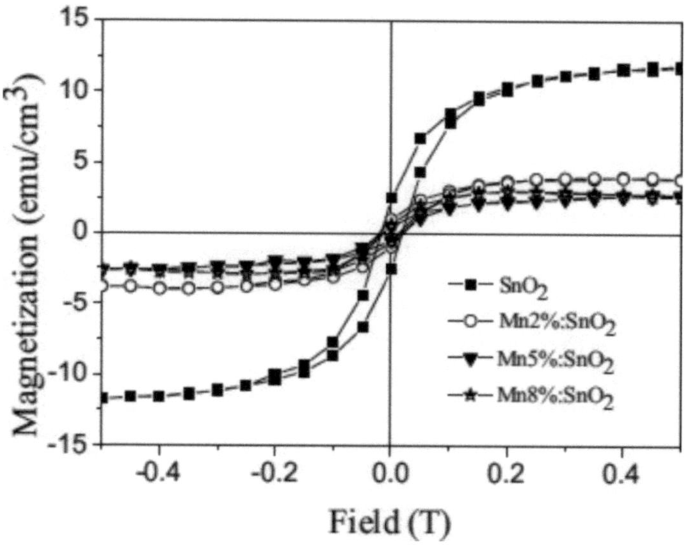

Figure 5.
RTFM for undoped and Mn doped tin oxide thin films, room temperature ferromagnetism.

synthesized by sol-gel method show that dopants and electronic cloud interactions play a significant role in establishing FM [44]. The ferromagnetic property of Mn doped SnO_2 confirmed that BMP's overlapping, oxygen vacancies and F-center exchange interaction are the cause for the existence of ferromagnetic behavior in in pure and doped Mn doped SnO_2 materials [51]. The increment of Mn concentration lead to the decline of magnetic moment of the origin of ferromagnetism behavior in Mn-doped SnO_2 films is explained BMP and the average magnetic moment per Mn concentration decreases with increasing Mn content [44]. Overall, the origin of FM in Mn doped SnO_2 system is still controversial and there is no such exact cause FM in this material.

2.2 Ferromagnetism in Fe-doped SnO_2

2.2.1 Experimental

Rodrı et al. reported that Fe-SnO_2 thin films on the $LaAlO_3$ subtract were synthesized with PLD techniques. The doped SnO_2 target was synthesized with metallic Fe powders and SnO_2. The powders were mixed with a ball-mill for 2 minutes, then pressed uniaxially (200 MPa) into a disk and finally sintered at 1000°C [52]. The crystallographic structures of the prepared Fe doped SnO_2 were determined by X-ray diffraction (XRD) and the magnetic measurements were performed with superconducting quantum interference device (SQUID).

2.2.2 Result and discussion

There have been reports of Fe doped SnO_2 in which the ferromagnetic interactions between magnetic impurities mediated oxygen or free carriers in the Fe doped SnO_2 system responsible for forming FM. Similarly, defects in undoped SnO_2 semiconductors may contribute to the observed ferromagnetism [33, 53]. Similarly both undoped and Fe-doped SnO_2 thin films shows the observed FM property is due to oxygen vacancies near Fe increased the magnetic moment, the RTFM behavior observed in the SnO_2 film must be associated with the sample shape or to defects incorporated during film growth and, part of the magnetism observed in SnO_2 as shown in figure [52]. The results from Fe-doped SnO_2 powders prepared by polymerized complex method confirms that the annealing temperature contributes to decline of magnetic saturation which is related to the defects rather than from dopant iron sites shown in **Figure 6a** below [54, 55] the existence of vacancies and defects in the grain boundaries and interfaces in Fe doped SnO_2 nanoparticles leads to decline the ferromagnetic behavior of system [56]. The decrease in Fe ion's magnetic moments, with their doping concentrations increasing, The super-exchange interaction may result in the interaction between neighboring TM-ions of the anti-ferromagnetic form, resulting in the observed decrease in the magnetic moment with increased concentration of TM as shown in **Figure 10** [57]. The work of the other group reporting shows that the lattice distortion induced by co-doping Fe-SnO_2 enhance ferromagnetic saturation magnetization of the compared with not co-doped one [26]. As shown in **Figure 7** oxygen vacancies has a great impact on the FM property of Fe doped SnO_2 [58]. The ferromagnetic behavior of Fe-doped SnO_2 thin films is caused by the coupling of ferric ions through an electron trapped in an oxygen vacancy [59]. In some cases the introduction of iron in semiconducting nano particles SnO_2 is responsible for appearance of paramagnetic behavior of the system that is due to weak antiferromagnetic interaction [60]. The decline of antiferromagnetic interaction in Fe-SnO_2 nano particles was reduced by the increment of Fe concentration [61]. Some study reported that the magnetic properties Fe-doped SnO_2 nano powders shows that the an increased Fe concentration leading to the reduction of oxygen-related vacancy changes magnetic property to paramagnetic

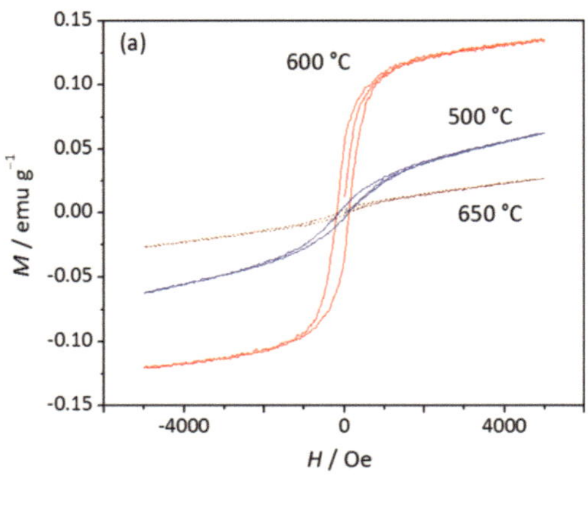

Figure 6.
Fe doped SnO_2 magnetic hysteresis annealed for certain hours at different temperatures [54].

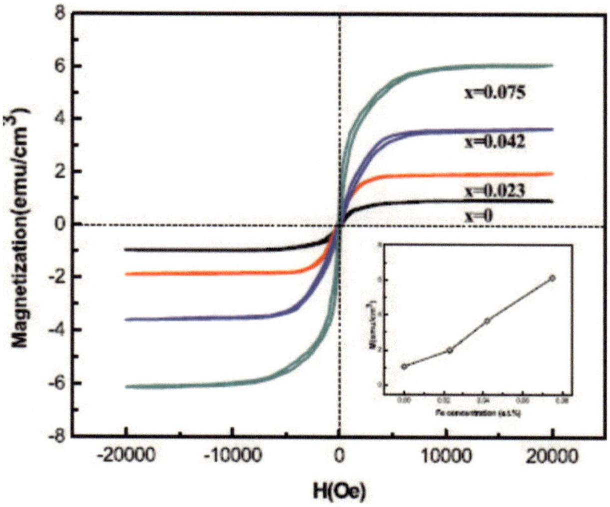

Figure 7.
Sn_{1-x} Fe_xO_2 film hysteresis loops with different concentration at 300 K. The inset is the difference in the concentration of Fe doping in saturation magnetization [58].

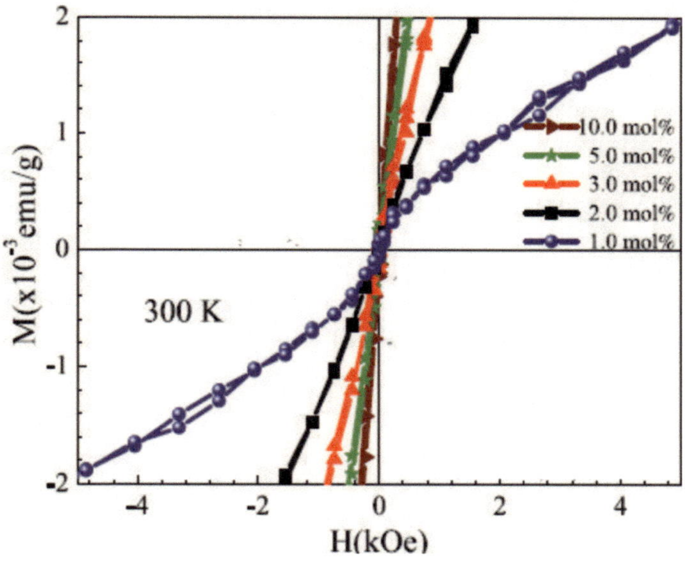

Figure 8.
RT magnetization versus magnetic field curves for the Fe-doped SnO_2 nano powders [62].

system as shown in **Figure 8** and the FM interactions is based on Bound Magnetic Polarons (BMPs) formation [62]. Sometimes the host systems SnO_2 and SnO doped with Fe during sample preparation can affect the observed magnetic properties of the system [63]. Another study confirms that changes in temperature play a role in the magnetic transition from paramagnetic to ferromagnetic behavior at ambient temperatures and low temperatures.

2.3 Ferromagnetism in Ni doped SnO_2

2.3.1 Experimental

As reported in the literature $SnNiO_2$ films were prepared by sol-gel method [64] the same procedure as [44]. The solution was prepared by dissolving $SnCl_4$ and

$NiCl_2{\cdot}6H_2O$ in distilled water and ethanol. For the film preparation, to get the thin films the solution was spin-coated on silicon substrate. As reported by [57] undoped and Ni doped SnO_2 prepared using a method of co-precipitation, the solution was prepared by dissolving $SnCl_4.5H_2O$ and $NiCl_2.5H_2O$ properly into de-ionized water. After the white precipitates were obtained, ammonium hydroxide (NH_4OH) was added with stirring to the solution. The resulting mixtures were washed with de-ionized water to remove unwanted ionic impurities that may develop during the process of synthesis. Such washed precipitates were dried in air and Ni doped SnO_2 powder products were eventually obtained.

Detail crystallographic structures of the prepared $SnNiO_2$ thin films and powders were carried out using XRD and the details of the magnetic properties were probed by vibrating sample magnetometer (VSM) measurements.

2.3.2 Result and discussion

According to recent experimental investigation of RTFM on Ni doped SnO_2 has made it important and promising materials for spintronics application [64–68]. The observed FM in these materials is linked to oxygen vacancy and structural defects of the materials [67]. In some cases, nano-crystalline Ni doped SnO_2 exhibits Paramagnetic character [68]. As shown in **Figure 9** the super-exchange interaction may result in an anti-ferromagnetic type interaction between neighboring TM-ions, resulting in a decline in magnetic moment with an increase in TM concentration [57]. In some studies the introduction of more Ni doping concentration leads for reduction of magnetic moment Ni ion of because the antiferromagnetic super-exchange interaction among closest neighbor in Ni^{2+} ions in Ni doped SnO_2 samples the BMP model can explain for RTFM on these systems on the other hand nickel (Ni) doped SnO_2 powder shows a substitution of Sn atom by Ni atom interstitially lead to the appearance of diamagnetic state [64]. Kuppan et al. shows that oxygen vacancy around magnetic impurity plays a major role in establishing ferromagnetism in Ni doped SnO_2. Nevertheless, saturation magnetization slowly decreases with a persistent rise in Ni doping concentration [68]. Thus we strongly feels that

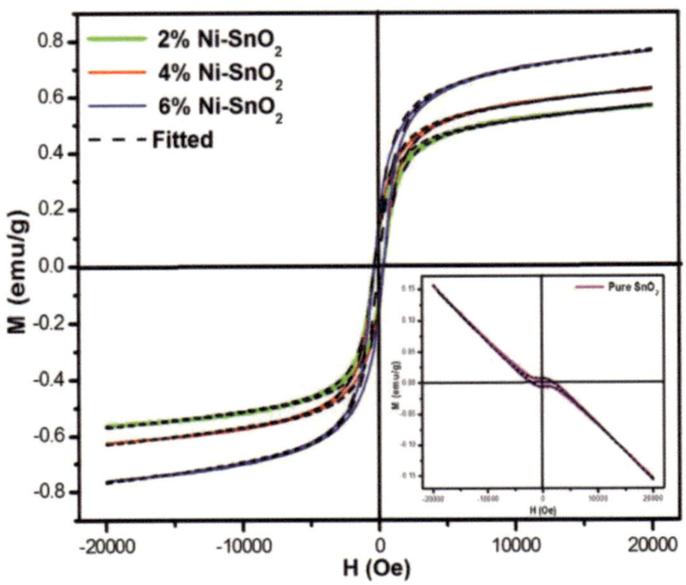

Figure 9.
The RTFM hysteresis of $SnFeO_2$ *and* $SnNiO_2$ *at different doping concentration [57].*

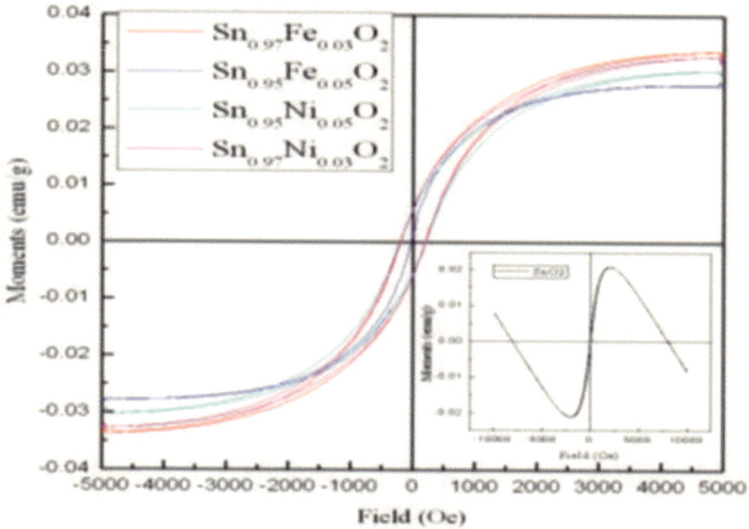

Figure 10.
Room temperature magnetization for pure and Ni doped SnO_2 NPs [69].

the oxygen vacancy and or defects in the Ni doped SnO_2 system. Similarly a report from Ni-doped SnO_2 nanoparticles synthesized by a polymer precursor method demonstrated that doping small amount of Ni doping concentration can push defect-related FM while introducing high Ni concentration favors the paramagnetic phase stabilization [70]. Similarly oxygen vacancy and defects on Ni doped SnO_2 thin film contribute for the formation RTFM [71]. As shown in **Figure 10** some studies have confirmed that Ni ions doping creates numerous defects or oxygen vacancies in SnO_2 nanoparticles in order to introduce RTFM in SnO_2 nanoparticles [69]. Some reported that substrates on thin film deposition have a strong impact on the magnetic moment of these material and the result confirmed that that FM in the films is as result of the doped matrix grown in different substrates [65]. In some cases the decrease of magnetic moment of per Ni ion is observed with the introduction of more dopants Ni ions that is associated with antiferromagnetic super-exchange interaction between in Ni ions in the system [64, 72]. Some reported that the mechanism of the observed FM in nickel (Ni) doped tin oxide thin films can be explained in bound magnetic polaron (BMP) mode [73].

3. Conclusion

Most of the results reported in the review shows that the perfect mechanism of induced FM in Mn, Fe and Ni doped SnO_2 is related to the intrinsic nature of the material itself, especially oxygen vacancies and defects of the crystal formed during sample preparation and doping magnetic impurity influence the magnetism of the systems. Even though, a different type of FM is reported. However, the different reported results are contradictory with each other and further research is needed to bring new solution for the contradiction idea behind FM.

Author details

Yared Worku[1], Dipti Ranjan Sahu[2*] and Vijaya Vallabhapurapu Srinivasu[1]

1 Department of Physics, University of South Africa (UNISA), Johannesburg, South Africa

2 Department of Natural and Applied Sciences, Namibia University of Science and Technology, Windhoek, Namibia

*Address all correspondence to: diptirs@yahoo.com

References

[1] Hummel RE. Electronic properties of materials. Springer Science & Business Media, Switzerland; 2011

[2] Fert A. The present and the future of spintronics. Thin Solid Films. 2008;**517**(1):2-5

[3] Das S. Spintronics a new class of device based on electron spin, rather than on charge, microelectronics may yield the next generation of microelectronics. American Scientist. 2013;**89**(6):516-523

[4] Bandyopadhyay S, Cahay M. Introduction to Spintronics. UK: CRC Press; 2015

[5] Akdogan N. Origin of ferromagnetism in oxide-based diluted magnetic semiconductors. Tese (Doutorado). Alemanha: Ruhr-Universität Bochum; 2008

[6] Fusil S, Garcia V, Barth A. Magnetoelectric devices for spintronicsr. Annual Review of Materials Research. 2014;**44**:91-116

[7] Furdyna JK. Diluted magnetic semiconductors. Journal of Applied Physics. 1988;**64**(4):R29-R64

[8] Kacman P. Spin interactions in diluted magnetic semiconductors and magnetic semiconductor structures. Semiconductor Science and Technology. 2001;**16**(4):R25

[9] Pearton SJ et al. Effects of defects and doping on wide band gap ferromagnetic semiconductors. Physica B: Condensed Matter. 2003;**342**:39-47

[10] Prestgard MC, Siegel GP, Tiwari A. Oxides for spintronics: A review of engineered materials for spin injection. Advanced Materials Letters. 2014;**5**(5):242-247

[11] Bhuvana S, Ramalingam HB, Thilakavathi G, Vadivel K. Structural, optical and magnetic properties of (Ni–Mn) co-doped tin oxide nanoparticles. Materials Technology. 2017;**7857**:1-5

[12] Pearton SJ, Abernathy CR, Overberg ME, Thaler GT, Norton DP. Wide band gap ferromagnetic semiconductors and oxides. Applied Phsiycs Reviews: Focused Review. 2003;**2013**(1)

[13] Coey JMD. Dilute magnetic oxides. 2007;**10**(2006):83-92

[14] Minami T. New n-type transparent conducting oxides. MRS Bulletin. 2000;**25**(8):38-44

[15] Fukumura T, Yamada Y, Toyosaki H, Hasegawa T. Exploration of oxide-based diluted magnetic semiconductors toward transparent spintronics. Applied Surface Science. 2004;**223**:62-67

[16] Fukumura T, Masashi K. Magnetic oxide semiconductors: On the high-temperature ferromagnetism in TiO_2-and ZnO-based compounds. Functional Metal Oxides: New Science and Novel Applications. 2013:89-131

[17] Philip J, Punnoose A, Kim BI, Reddy KM, Layne S, Holmes JO, et al. Carrier-controlled ferromagnetism in transparent oxide semiconductors. Nature Materials. 2006;**5**(4):298-304

[18] Prins MWJ et al. A ferroelectric transparent thin-film transistor. Applied Physics Letters. 1996;**3650**:1-4

[19] Sahana MB, Sudakar C, Setzler G, Dixit A, Thakur JS. Bandgap engineering by tuning particle size and crystallinity of SnO_2-Fe_2O_3 nanocrystalline composite thin films. Applied Physics Letters. 2012;**231909**(2008):2-5

[20] Batzill M, Diebold U. The surface and materials science of tin oxide. Progress in Surface Science. 2005;**79**:47-154

[21] Oshima M, Yoshino K. Structural and Electronic Structure of SnO_2 by the First-Principle Study. Materials Science Forum. 2012;**725**:265-268

[22] Nicoletti S, Dori L, Cardinali GC, Parisini A. Gas sensors for air quality monitoring: Realisation and characterisation of undoped and noble metal-doped SnO_2 thin sensing films deposited by the pulsed laser ablation. Sensors and Actuators B: Chemical. 1999;**60**:90-96

[23] Al-gaashani R, Radiman S, Tabet N, Daud AR. Optical properties of SnO_2 nanostructures prepared via one-step thermal decomposition of tin (II) chloride dihydrate. Materials Science and Engineering B. 2012;**177**(6):462-470

[24] Prellier W, Fouchet A, Mercey B. Oxide-diluted magnetic semiconductors: A review of the experimental status. Journal of Physics: Condensed Matter. 2003;**15**(37):R1583

[25] Physics N, Flatt ME. Challenges for semiconductor spintronics; 2014

[26] Nomura K et al. Magnetic properties of Fe and Co codoped SnO_2 prepared by sol-gel method. Journal of Applied Physics. 2013;**083901**(2011):1-5

[27] Davis KB, Stringari S, Wieman CE. Zener Model Description of Ferromagnetism in Zinc-Blende. Science. 2000;**287**:1019-1023

[28] Alekseevich AA, Vakhitov IR, Khaibullin RI, Tagirov LR. On the nature of ferromagnetism in oxide semiconductors doped with 3d-elements. Journal of Physics: Conference Series. 2012;**394**(1):012018

[29] Jürgen K, Lin HH, MacDonald AH. Theory of diluted magnetic semiconductor ferromagnetism. Physical Review Letters. 2000;**84**(24):5628

[30] Park JH, Kim MG, Jang HM, Ryu S, Kim YM. Co-metal clustering as the origin of ferromagnetism in Co-doped ZnO thin films. Applied Physics Letters. 2004;**84**(8):1338-1340

[31] Van Komen C et al. Structure-magnetic property relationship in transition metal (M = V, Cr, Mn, Fe, Co, Ni) doped SnO_2 nanoparticles. Journal of Applied Physics. 2010;**141**:2006-2009

[32] Hong NH. Ferromagnetism in transition-metal-doped TiO_2 thin films. Physical Review B. 2004;**70**:1-6

[33] Hong NH, Sakai J, Huong NT, Poirot N, Ruyter A. Role of defects in tuning ferromagnetism in diluted magnetic oxide thin films. Physical Review B; 2005;**72**(4):045336

[34] Seghier DÃ, Gislason HP. On the oxygen vacancy in Co-doped ZnO thin films. Physica B: Physics of Condensed Matter. 2009;**404**(23-24):4800-4802

[35] Vlasenko LS. Magnetic resonance studies of intrinsic defects in ZnO: Oxygen vacancy. Applied Magnetic Resonance. 2010;**39**:103-111

[36] Sato K, Katayama-Yoshida H. Material design for transparent ferromagnets with ZnO-based magnetic semiconductors. Japanese Journal of Applied Physics. 2000;**39**(6B):L555

[37] Weng H, Yang X, Dong J, Mizuseki H, Kawasaki M, Kawazoe Y. Electronic structure and optical properties of the Co-doped anatase TiO_2 studied from first principles. Physical Review B. 2004;**69**(12):125219

[38] Rahman G, García-suárez VM, Hong SC. Vacancy-induced magnetism in SnO_2: A density functional study. Physical Review B: Condensed Matter. 2008;**78**:1-5

[39] Fitzgerald CB, Venkatesan M, Dorneles LS, Gunning R, Stamenov P, Coey JMD. Magnetism in dilute magnetic oxide thin films based on SnO_2. Physical Review B. 2006;**74**:1-10

[40] Ogale SB et al. High temperature ferromagnetism with a giant magnetic moment in transparent Co-doped SnO_2. Physical Review Letters. 2003;**91**:1-4

[41] Coey JMD, Douvalis AP, Fitzgerald CB, Venkatesan M. Ferromagnetism in Fe-doped SnO_2 thin films. Applied Physics Letters;**84**:1332-1334

[42] Punnoose A et al. Development of high-temperature ferromagnetism in SnO_2 and paramagnetism in SnO by Fe doping. Physical Review B. 2005;**72**

[43] Nomura K. Magnetic properties and oxygen defects of dilute metal doped tin oxide based semiconductor. Croatica Chemica Acta. 2016;**88**(4):579-590

[44] Xiao Y et al. Room temperature ferromagnetism of Mn-doped SnO_2 thin films fabricated by sol-gel method. Applied Surface Science. 2008;**254**(22):7459-7463

[45] Fitzgerald CB et al. SnO_2 doped with Mn, Fe or Co: Room temperature dilute magnetic semiconductors. Journal of Applied Physics. 2011;**7390**(2004):25-28

[46] Tian ZM et al. Structure and magnetic properties in Mn doped SnO_2 nanoparticles synthesized by chemical co-precipitation method. Journal of Alloys and Compounds. 2008;**466**:26-30

[47] Kimura H, Fukumura T, Kawasaki M, Inaba K, Hasegawa T, Koinuma H. Rutile-type oxide-diluted magnetic semiconductor: Mn-doped SnO_2. Applied Physics Letters. 2005;**94**(2002):78-81

[48] Hong NH, Poirot N, Sakai J, Sno M. Ferromagnetism observed in pristine SnO_2 thin films. Physical Review B: Condensed Matter. 2008;**77**:033205

[49] Ahmed SA. Room-temperature ferromagnetism in pure and Mn doped SnO_2 powders. Solid State Communications. 2010;**150**(43-44):2190-2193

[50] Ahmed SA, Mohamed SH. Room temperature ferromagnetism behavior of $Sn_{1-x}Mn_xO_2$ powders. Journal of Magnetism and Magnetic Materials. 2012;**324**(5):812-817

[51] Ahmad N, Khan S, Mohsin M, Ansari N. Optical, dielectric and magnetic properties of Mn doped SnO_2 diluted magnetic semiconductors. Ceramics International. 2018;**44**(13): 15972-15980

[52] Rodrı CE, Errico L, Golmar F, Navarro AMM, Cabrera AF. The role of the dopant in the magnetism of Fe-doped SnO_2 films. 2007;**316**:219-222

[53] Nomura K, Barrero CA, Sakuma J, Takeda M. Room-temperature ferromagnetism of sol-gel-synthesized $Sn_{1-x}57Fe_xO_2$ powders. Physical Review B. 2007;**75**:1-13

[54] Nomura K. Magnetic properties and oxygen defects of dilute metal doped tin oxide based semiconductor. Croatica Chemica Acta. 2015;**88**(4):579-590

[55] Sakuma J, Nomura K, Barrero C, Takeda M, Science F. Mössbauer studies and magnetic properties of SnO_2 doped. Thin Solid Films. 2007;**515**:8653-8655

[56] Mishra AK, Sinha TP, Bandyopadhyay S, Das D. Structural and magnetic properties of nanocrystalline Fe-doped SnO_2.

Materials Chemistry and Physics. 2011;**125**(1-2):252-256

[57] Sharma A, Varshney M, Kumar S, Verma KD, Kumar R. Magnetic properties of Fe and Ni doped SnO_2 nanoparticles. Nanomaterials and Nanotechnology. 2011;**1**(1):6

[58] Fu Y, Sun N, Feng L, Wen S, An Y, Liu J. Local structure and magnetic properties of Fe-doped SnO_2 films. Journal of Alloys and Compounds. 2017;**698**:863-867

[59] Coey JMD, Douvalis AP, Fitzgerald CB, Venkatesan M. Ferromagnetism in Fe-doped SnO_2 thin films. Applied Physics Letters. 2004;**84**(8):1332-1334

[60] Sambasivam S, Chun B, Lin JG. Intrinsic magnetism in Fe doped SnO_2 nanoparticles. Journal of Solid State Chemistry. 2011;**184**:199-203

[61] Adhikari R, Das AK, Karmakar D. Structure and Magnetism of Fe-Doped SnO_2 Nanoparticles. Physical Review B: Condensed Matter. 2008;**78**:024404

[62] Aragón FH, Coaquira JAH, Gonzalez I, Nagamine LCCM, Macedo WAA, Morais PC. Fe doping effect on the structural, magnetic and surface properties of SnO_2 nanoparticles prepared by a polymer precursor method. Journal of Physics D: Applied Physics. 2016;**49**(15):155002

[63] Punnoose A, Hays J, Thurber A, Engelhard ME, Ravi K et al. Development of high-temperature ferromagnetism in SnO_2 and paramagnetism in SnO by Fe doping. Physical Review B. 2005;**72**(5):054402

[64] Zhang J, Yun Q, Wang Q. Room temperature ferromagnetism of Ni-doped SnO_2 system. Modern Applied Science. 2010;**4**(11):124-130

[65] Hong NH, Ruyter A, Prellier W, Sakai J, Huong NT. Magnetism in Ni-doped SnO_2 thin films. Journal of Physics: Condensed Matter. 2005;**17**(41):6533

[66] Aragón, FH, Coaquira JAH, Hidalgo P, Brito SLM, Gouvêa D, Castro RHR. Structural and magnetic properties of pure and nickel doped SnO_2 nanoparticles. Journal of Physics: Condensed Matter. 2010;**22**(49):496003

[67] Wang XL, Dai ZX, Zeng Z. Search for ferromagnetism in SnO_2 doped with transition metals (V, Mn, Fe, and Co). Vol. 0452142008

[68] Kuppan M, Kaleemulla S, Rao NM, Krishna NS, Begam MR, Shobana M. Structural and magnetic properties of Ni doped SnO_2 powders. Advances in Condensed Matter Physics. 2014;**2014**

[69] Ali T. Enhanced room temperature ferromagnetism in Ni doped SnO_2 nanoparticles: A comprehensive study. Journal of Applied Physics. 2017;**122**:083906

[70] Aragón FH, Coaquira JAH, Hidalgo P, Brito SLM, Gouvêa D, Castro RHR. Structural and magnetic properties of pure and nickel doped SnO_2 nanoparticles. Journal of Physics: Condensed Matter. 2010;**22**(49):496003

[71] Pascariu P, et al. Structural, optical and magnetic properties of Ni doped SnO_2 nanoparticles. Journal of Alloys and Compounds. 2016;**668**:65-72

[72] Sharma M, Kumar S, Alvi PA. Electronic structure and magnetic properties of Ni-doped SnO_2 thin films. AIP Conference Proceedings. 2018;**120010**:2-5

[73] Abdi MH, Ibrahim NB, Baqiah H, Halim SA. Structural, electrical and magnetic characterization of nickel-doped tin oxide lm by a sol-gel method. Scientia Iranica. 2014;**21**:2459-2467

Section 4

Magnetic Levitation

Chapter 8

Magnetic Levitation Based Applications in Bioscience

Fatih Ozefe and Ahu Arslan Yildiz

Abstract

Contactless manipulation of small objects, such as micro−/nanoparticles, biological entities, and even cells is required in varied applications in biosciences. Magnetic levitation (MagLev) is a new-generation methodology to achieve contactless magnetic manipulation of objects. Lately, magnetic levitation methodology has been utilized in several applications in bioscience, such as biosensors, diagnostics and tissue engineering. Magnetic levitation enables separation or positioning of objects in three-dimensional (3D) space based on their density features. Therefore, density-based separation assays utilizing magnetic levitation for biosensing or diagnostic purposes are developed recently. Specific particles or cells, which are markers of any disease, could be detected by sorting them based on density differences through magnetic levitation. On the other hand, tissue engineering studies and production of self-assembled 3D cell culture structures are carried out by magnetic levitation, where cells are magnetically positioned while allowing cell-cell interaction resulting in 3D cell culture formation. Lately, magnetic levitation methodologies received more interest in the field of bioscience due to advantages about the efficiency and cost. This contribution broadly summarizes recent efforts in magnetic levitation techniques that are mainly applied in diagnostics and tissue engineering.

Keywords: magnetic levitation, tissue engineering, diagnostic tools, biosensors, density-based assay, contactless manipulation

1. Introduction

Mimicking the microenvironment of biological systems is crucial, especially for diagnostic and tissue engineering purposes. There are several contact-free manipulation methodologies [1] that mimic and control the microenvironments of biological systems, such as magnetophoresis [2], acoustophoresis [3], electrophoresis [4], and thermophoresis [5]. Magnetophoresis is a method that provides contact-free manipulation of particles in a magnetic field. There is no additional equipment required instead of permanent magnets or electromagnets for magnetophoresis; however, sound waves, electrical source, and heat source are required for acoustophoresis, electrophoresis, and thermophoresis, respectively [6].

Magnetophoresis is a contactless manipulation method, which provides the manipulation of particles in the magnetic field provided by either permanent magnets or electromagnets. During manipulation of particles in a magnetic field gradient or/and in a magnetized medium, neither pH nor temperature of sample is affected [7]. There are two different types of magnetophoresis, such as positive and negative magnetophoresis (**Figure 1A, B**). In the positive magnetophoresis, particles that have magnetic properties migrate within nonmagnetic (diamagnetic) medium. On the other hand, particles that do not have magnetic properties migrate within paramagnetic medium in negative magnetophoresis [8, 9]. The migration of particles in both types of magnetophoresis depends on the magnetic susceptibility differences between particle and medium.

Magnetic levitation (MagLev) technique, which works with the principle of negative magnetophoresis, manipulates the diamagnetic particles in paramagnetic medium by providing antigravity conditions (**Figure 1C**) [10]. The diamagnetic particles that are suspended in a paramagnetic medium are positioned at specific height called levitation height depending on their densities when the external magnetic field is applied. According to magnetic levitation principle, diamagnetic particles are specifically positioned depending on their densities by balancing all forces on particles, which are gravitational force (from gravity) and magnetic buoyant force (from magnetic field).

In magnetic levitation systems, biological entities (as diamagnetic particles) are also levitated and manipulated in a three-dimensional (3D) space as well as nonbiological particles in the paramagnetic medium [11]. The paramagnetic medium is an important parameter for magnetic levitation-based approaches, because magnetic susceptibility differences between paramagnetic medium ($\chi > 0$) and diamagnetic particle ($\chi < 0$) provide magnetic buoyancy force on diamagnetic particles in the presence of external magnetic field [12]. There are different paramagnetic mediums that are used in magnetic levitation-based approaches, such as ferrofluids [13] and paramagnetic salt solutions [2]. Ferrofluids are the suspension of maghemite (Fe_2O_3) or magnetite (Fe_3O_4) that has higher magnetic susceptibility than paramagnetic salt solutions; however, observation of the samples is limited because of their opaque nature [14]. Manganese (II) chloride ($MnCl_2$) and diethylenetriamine-pentaacetic acid (Gd-DTPA) are generally chosen and used paramagnetic agents in magnetic levitation; however, they are used in high concentration to effectively levitate diamagnetic particles because of their low magnetic susceptibilies. The use of paramagnetic salts in high concentration is the main reason for toxicity-based limitations in bioapplications of magnetic levitation technology [2].

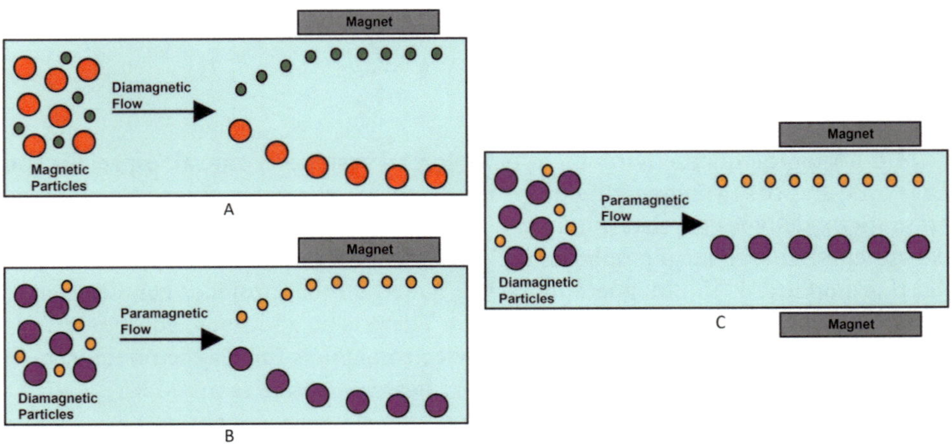

Figure 1.
Migration of particles with magnetophoresis. (A) Positive magnetophoresis, which means the migration of magnetic particles in diamagnetic fluid. (B) Negative magnetophoresis, which means the migration of diamagnetic particles in paramagnetic fluid. (C) Migration of diamagnetic particles in paramagnetic fluid via magnetic levitation principle.

1.1 History and theory of magnetic levitation

Magnetic levitation concept appeared within the study named as "An Absolute Micromanometer Using Diamagnetic Levitation" at the end of 1960s. In this study, the diamagnetically levitated ultramicromanometer was described where friction-free suspension was produced via magnetic induction for graphite disk to measure absolute pressure down to 10^{-10} Torr [15]. Later, density-based separation was carried out for minerals [16] and metals [17] by magnetic levitation principle. In another study, improvement of magnetic levitation system has been studied to measure the density differences of liquids and solids [18–20]. On the other hand, magnetic levitation-based approaches also appeared in tissue engineering applications. The magnetic levitation techniques were also used to mimic 3D cellular microenvironment and to form 3D cellular structures by guiding the cells [21–26].

As depicted in **Figure 2**, the diamagnetic particles in paramagnetic medium are density-dependently positioned at specific height, in which each force (gravitational, magnetic buoyant forces, and steric interactions) on diamagnetic particle (**Figure 2B**) is equalized under magnetic field produced by permanent magnets [Eqs. (1) and (4)] [27–29]. The magnetic and gravitational forces act on diamagnetic particles and cause the particles to either float or sink in paramagnetic solution (**Figure 2C**) [1].

At balanced height:

$$\vec{F}_g + \vec{F}_m = 0 \tag{1}$$

$$\vec{F}_g = \left(\rho_p - \rho_m\right) V\vec{g} \tag{2}$$

$$\vec{F}_m = \frac{\left(X_p - X_m\right)}{\mu_0} V\left(\vec{B}.\vec{\nabla}\right)\vec{B} \tag{3}$$

$$\vec{F}_g + \vec{F}_m = \left(\rho_p - \rho_m\right) V\vec{g} + \frac{\left(X_p - X_m\right)}{\mu_0} V\left(\vec{B}.\vec{\nabla}\right)\vec{B} = 0 \tag{4}$$

In Eq. (4), the gravitational $\left(\vec{F}_g\right)$ and magnetic $\left(\vec{F}_m\right)$ forces on levitating particles are balanced. Here, ρ_p and ρ_m refer to the density of levitating particle and paramagnetic medium (kg.m^{-3}), respectively; and X_p and X_m represent the

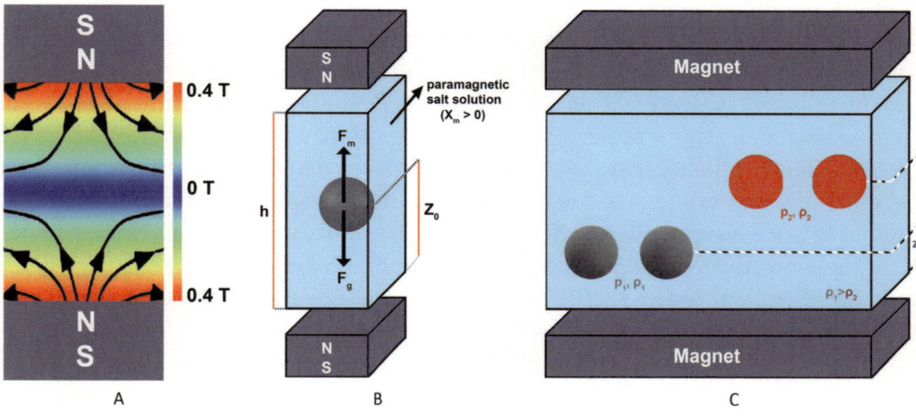

Figure 2.
The principle of magnetic levitation. (A) Magnetic field between the magnets, which is oriented in anti-Helmholtz configuration. (B) the forces on levitating objects within the magnetic levitation systems. (C) the alignment of levitating objects at specific levitation heights depending on their densities.

magnetic susceptibility of levitating particle and paramagnetic medium (unitless), respectively. V is the volume of levitating particle (m^3); g is the gravitational constant ($9.81\ m.s^{-2}$); μ_0 is the permeability of free space ($1.26 \times 10^{-6}\ kg.s^{-2}.A^{-2}$); and B is the magnitude of applied magnetic field (T).

$$Z_0 = \left[\frac{g\mu_0 h^2}{(X_s - X_m)4B_0^2}\right]\rho_p + \left[\frac{h}{2} - \frac{\rho_m g\mu_0 h^2}{(X_s - X_m)4B_0^2}\right] \tag{5}$$

According to Eq. (4), the levitation height of levitating particles, which is shown as Z_0 (m) (**Figure 2B**), could be mathematically determined in certain space between magnets, which is shown as h (m) Eq. (5) [28].

2. Magnetic levitation approaches in biosciences

2.1 Magnetic levitation technology for biosensors and diagnostics

Magnetic levitation-based technologies were recently developed for positioning and sorting of diamagnetic particles, which can be either biological or nonbiological materials to be used in biosensing and diagnostic purposes. The magnetic levitation systems in diagnostic field are mainly composed of strong magnets and optical components. N52-grade NdFeB magnets are oriented in anti-Helmholtz configuration where same poles face toward each other. N52-grade NdFeB magnets produce magnetic field around 0.4 T, which causes the magnetic buoyant force on diamagnetic particles suspended in paramagnetic medium [1].

In the earliest version (**Figure 3A**), magnetic levitation system is fabricated from polydimethylsiloxane (PDMS) material, which is combined with NdFeB magnets. These systems had capability to dynamically separate polymeric diamagnetic particles suspended in paramagnetic salt solution ($GdCl_3$) by differentiation of levitation height depending on their density differences. While the diamagnetic particles with highest densities were moved through bottom part of the channel, the diamagnetic particles with lowest densities were moved through upper part of the channel [29]. After that, simple cuvette was integrated into magnetic levitation system instead of PDMS holder to analyze the quality of chemical reactions and also to statically measure the density of diamagnetic solid particles. As shown in **Figure 3B**, magnetic levitation technique was utilized to monitor chemical reactions by observing the levitation height changes of polymeric beads. The decrease in the levitation heights was observed depending on increasing the density of polymeric beads during solid-phase reaction [28]. Also, the density determination of materials with unknown densities was performed by using density versus levitation height curve, which is produced from materials with known densities [18]. To improve measurable density range that could be measured by magnetic levitation system, "Tilted MagLev" system (**Figure 3C**) was developed. In such system, the range of measurable densities was expanded around 15-fold; however, the sensitivity of "Tilted MagLev" was limited [19]. To overcome the sensitivity limitations in "Tilted MagLev," high-sensitivity magnetic levitation setup was developed by rotating "Tilted MagLev." As shown in **Figure 3D**, density measurement could be performed with resolution down to 10^{-6} g/ml, which means the rotated configuration improved the sensitivity 100-fold over previous [20].

On the other hand, a magnetic levitation platform has been developed for analysis of proteins' binding event. The diamagnetic polymeric beads were covalently functionalized with the derivatives of benzenesulfonamide to be used for capturing

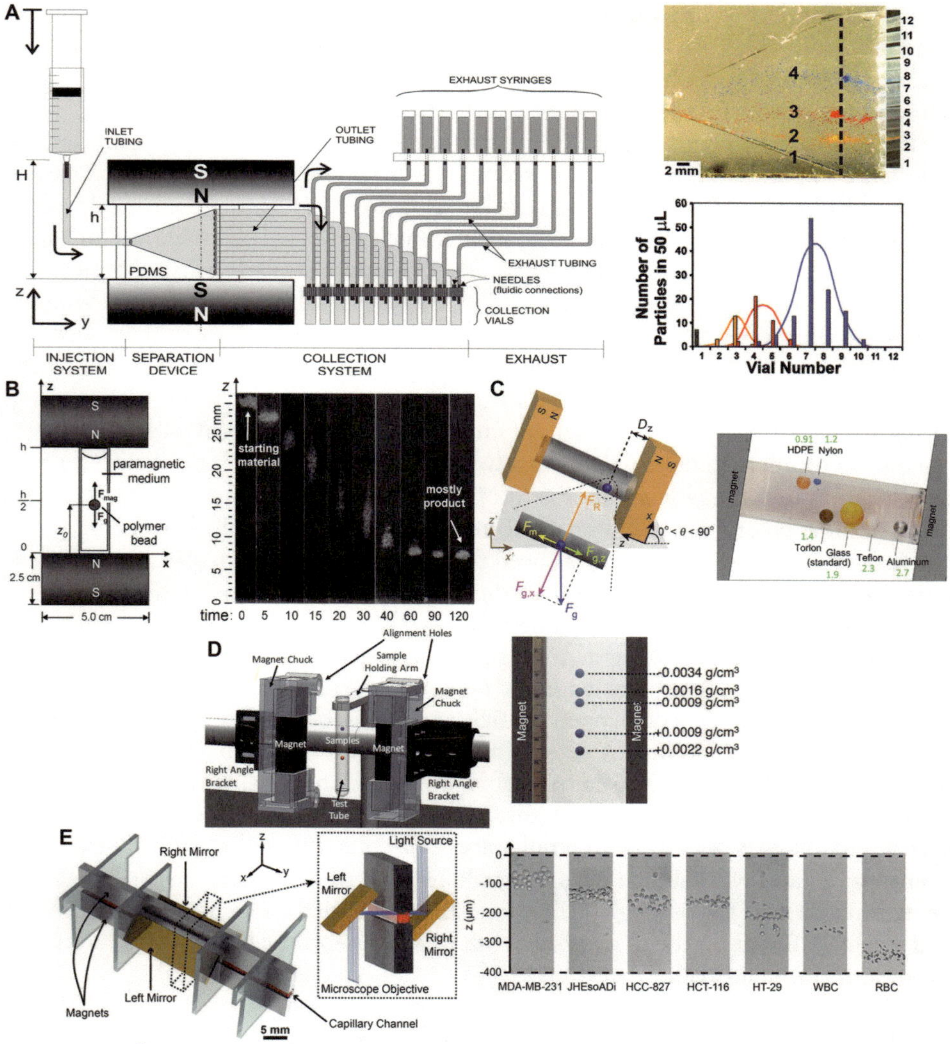

Figure 3.
Development and improvements in the magnetic levitation systems used in diagnostic applications. (A) the platform that had injection system, separation device, collection system, and exhaust was used in density-dependent separation. Reprinted with permission from Ref. [29], Copyright 2007 American Chemical Society. (B) the magnetic levitation device that contained capillary tube levitates the particles. Reprinted with permission from Ref. [28], Copyright 2008 American Chemical Society. (C) Tilted MagLev that provides the particles in broad range of densities. Reprinted with permission from Ref. [19], Copyright 2016 American Chemical Society. (D) The MagLev device that provides high-sensitive separation of particles depending on their densities. Reprinted with permission from Ref. [20], copyright 2016 American Chemical Society. (E) The magnetic levitation system that provides the analysis of cells via microscope. Reproduced with permission from Ref. [11], Copyright 2015, proceedings of the National Academy of Sciences USA.

BCA. After capturing protein in solution containing disodium gadolinium (III) diethylenetriamine pentaacetic acid, Gd(DTPA), the changes on the levitation heights of ligand-coated diamagnetic particles were analyzed in magnetic levitation systems. The levitation height of ligand-coated diamagnetic particles changed due to protein capture. Depending on levitation height change of ligand-coated diamagnetic particles, the interaction between ligand and protein, and furthermore protein concentration was determined in the sample [30]. Later, magnetic levitation system was improved for hepatitis C detection. In this system, magnetic levitation principle was combined with enzyme-linked immunosorbent assay (*ELISA*), called density-linked immunosorbent assay (*DeLISA*). The surface of diamagnetic polymeric beads was functionalized with HCV-NS3 protein, and levitation heights were determined before and after anti-HCV NS3 binding. Analyzing the levitation height differences, which are caused by interaction between ligand and protein, the detection of hepatitis C was carried out via *DeLISA* [31].

Especially for diagnostic purposes, a new-generation magnetic levitation device, which is compatible with light microscope, was designed and developed to analyze micro-sized particles, such as polymeric beads and cells (**Figure 3E**). This device is composed of two N52-grade NdFeB magnets, microcapillary channel, mirrors and poly(methyl methacrylate) (PMMA) holder. The microcapillary glass channel was integrated in between magnets (oriented in anti-Helmholtz configuration) where all system components were assembled within PMMA holder. The mirrors were placed in 45° angle to reflect light through microcapillary channel for providing the visualization of micro-sized particles within magnetic levitation system. As shown in **Figure 3E**, the developed system was highly sensitive for density-dependent separation of micro-sized particles and was used to profile densities of different cells, such as breast adenocarcinoma, esophageal adenocarcinoma, colorectal adenocarcinoma, colorectal carcinoma, and red and white blood cells [11]. Later, similar magnetic levitation system was used for diagnosis of malaria-infected red blood cells and sickle cells by analyzing their density-dependent levitation heights [32, 33]. Previously mentioned magnetic levitation systems [11, 32, 33] require highly sophisticated analyzing instruments such as light microscope that causes high cost. To reduce that cost, magnetic levitation system was combined with smartphone to analyze micro-sized particles and even cells. Smartphone-assisted systems allow levitation height determination of micro-sized particles in microcapillary channel by the camera of smartphone and lens, which are used for focusing [34–36]. One of these systems, namely "i-LEV," was used for density-dependent analysis of cells from whole blood via smartphone. In addition to counting red and white blood cells in whole blood, "i-LEV" could be used in the detection and analysis of single cells where there is a difference between healthy and sick cell. Based on the density-dependent alignment of cells, "i-LEV" was used for monitoring sickle cell disease [35].

2.2 Magnetic levitation technology for tissue engineering

Magnetic levitation systems have been also specialized to provide contactless manipulation of cells for the production of 3D cellular structures in tissue engineering applications. In the earliest version, magnetic force-based tissue-engineering (MagTE) setup was developed, which utilizes magnetic cationic liposome (MCL)-guided cells for the formation of 3D string or ring-shaped tissue structures under magnetic field provided by permanent magnets. MCLs in the culture medium were uptaken by myoblast C2C12 cells, and then labelled myoblast C2C12 cells were manipulated by inducing the magnetic field provided by permanent NdFeB magnets. The oriented skeletal muscle tissues, string-like and ring-like assemblies, were obtained [21]. Later, 3D bioassembler, which is M13 phage-based hydrogel containing magnetic iron oxide (MIO) and gold nanoparticles (Au/NP), was utilized for the formation of 3D cellular structures via magnetic levitation technology. The cells in magnetic levitation system interacted with each other and formed 3D clusters after 30 min right after cells were levitated. Varied 3D cellular structures were obtained by incubating clustered cells for 24 h. The different cell types, human glioblastoma cells and human astrocytes, were also 3D cocultured [22]. In another study, poly-L-lysine cross-linked MIO/AuNP hydrogel, termed as NanoShuttle, was developed and used in magnetic levitation system to form 3D cellular structures in 96-well plate from varied cell types, for example human embryonic kidney cells, mouse fibroblast cells, human mammary epithelial, human umbilical vein endothelial cell, and human hepatocyte cells. However, the morphologies of 3D cellular structures were different for each cell line depending on their cellular properties [25]. Also, the same magnetic levitation system was utilized for the formation of multicellular 3D cellular structures, called organoids. The bronchiole wall was formed by sequential layered assembly technique. Epithelial cells (EpiCs), smooth muscle cells (SMCs), pulmonary fibroblasts (PFs), and pulmonary endothelial cells (PEC) were individually grown and levitated for 4 h. Later, levitated cell lines were sequentially taken into bronchiole coculture as EpiCs, SMCs, PFs, and PECs [23]. The same procedure was also applied to form the aortic valve coculture formation from valvular interstitial cells (VICs) and endothelial cells (VECs) [24] and heterogeneous breast tumors from various breast cancer cells and fibroblast [26]. On the other hand, the generation of organoids, called as adipospheres, by the culturing of 3 T3-L1 preadipocyte cells and bEND.3 endothelial cells with magnetic levitation technique has been accomplished [37].

Recently, a new magnetic levitation system, which uses only paramagnetic salt solution instead of NanoShuttles or any other magnetic biomaterial, was developed. Cells were encapsulated into methacrylated gelatin (GelMa) or polyethylene glycol dimethacrylate (PEDGA) hydrogels and they were levitated in the presence of Gd^{3+} solution. 3D cell cultures were achieved by the levitational self-assembly of those microstructures [38].

In tissue engineering applications, the above-mentioned magnetic levitation systems have limitations because of the scaffold or additional compounds were required for 3D cell culturing. The new-generation magnetic levitation system that allows scaffold-free cell culturing was developed to overcome with these limitations [39]. In such system, cells are three cultured in 3D microenvironment without any scaffold materials or additional requirements. The mouse fibroblast cells (NIH 3T3) and non-small-cell lung cancer cells (HCC827) were mixed with Gadolinium (III) (Gd^{3+}) chelate, which is a paramagnetic agent, before loading into developed magnetic levitation setup. With the help of paramagnetic agent, the cells were levitated and suspended at their specific levitation height in the presence of external magnetic field. By interacting with levitated cells, cells started to secrete their own extracellular matrixes, and cellular clusters were formed within the magnetic levitation setup. The 3D cellular spheroids were formed from low number of cells ($1x10^3$) in 48 h; however, the cellular strings were observed when a high number of cells were used ($1x10^5$) [39].

3. Summary and conclusion

Magnetic levitation-based technologies are new-generation systems that are used in biosensing, diagnostics, as well as tissue engineering. Density-dependent separation of particles provides high-speed analysis of biomolecules, even cells, with low cost in diagnostics. On the other hand, contactless manipulation of cells via magnetic force can also be provided by magnetic levitation systems. In such systems, cells are easily manipulated in the presence of magnetic field after they are magnetized with specialized molecules, such as ferrofluids and paramagnetic agents. By positioning the cells at a specific height in medium, where they can easily interact with each other, natural 3D microenvironment can be formed.

While the magnetic levitation-based systems provide faster, cost-effective, and easy-to-use diagnostic applications, at the same time, it is also used to form 3D cellular structures in tissue engineering field.

Acknowledgements

This work was partially supported by TUBITAK 118Z344 Grant and TUBA GEBIP 2019 award.

Conflict of interest

The authors declare no conflict of interest.

Author details

Fatih Ozefe and Ahu Arslan Yildiz*
Department of Bioengineering, Izmir Institute of Technology, Izmir, Turkey

*Address all correspondence to: ahuarslan@iyte.edu.tr

References

[1] Turker E, Arslan-Yildiz A. Recent advances in magnetic levitation: A biological approach from diagnostics to tissue engineering. ACS Biomaterials Science & Engineering. 2018;**4**(3):787-799. DOI: 10.1021/acsbiomaterials.7b00700

[2] Munaz A, Shiddiky MJA, Nguyen NT. Recent advances and current challenges in magnetophoresis based micro magnetofluidics. Biomicrofluidics. 2018;**12**(3)031501:1-23. DOI: 10.1063/1.5035388

[3] Chen X, Lam KH, Chen R. Acoustic levitation and manipulation by a high-frequency focused ring ultrasonic transducer. Applied Physics Letters. 2019;**114**(5);054103:1-5. DOI: 10.1063/1.5066615

[4] Yamada M, Harada M, Okada T. Electrophoretic levitation and focusing of particles by field switching. Analytical Sciences. 2009;**25**(2):279-284. DOI: 10.2116/analsci.25.279

[5] Piazza R. Thermophoresis: Moving particles with thermal gradients. Soft Matter. 2008;**4**(9):1740-1744. DOI: 10.1039/b805888c

[6] Zhu G, Nguyen NT. Particle sorting in microfluidic systems. Micro and Nanosystems. 2010;**2**(3):202-216. DOI: 10.2174/1876402911002030202

[7] Gao QH, Zhang WM, Zou HX. Label-free manipulation via the magneto-Archimedes effect: Fundamentals, methodology and applications. Materials Horizons. 2019;**6**(7):1359-1379. DOI: 10.1039/c8mh01616j

[8] Chen Y, Lin P, Huang PH. Rare cell isolation and analysis in microfluidics. Lab on a Chip. 2014;**14**(4):626-645. DOI: 10.1039/c3lc90136j

[9] Zhu GP, Hejiazan M, Huang X. Magnetophoresis of diamagnetic microparticles in a weak magnetic field. Lab on a Chip. 2014;**14**:4609-4615. DOI: 10.1039/c4lc00885e

[10] Ge S, Whitesides GM. "Axial" magnetic levitation using ring magnets enables Simple density-based analysis, separation, and manipulation. Analytical Chemistry. 2018;**90**:12239-12245. DOI: 10.1021/acs.analchem.8b03493

[11] Durmus NG, Tekin HC, Guven S. Magnetic levitation of single cells. Proceedings of the National Academy of Sciences of the United States of America. 2015;**112**(28):E3661-E3668. DOI: 10.1073/pnas.1509250112

[12] Catherall AT, Eaves L, King PJ. Floating gold in cryogenic oxygen. Nature. 2003;**422**(6932):579. DOI: 10.1038/422579a

[13] Rosensweig RE. Fluidmagnetic buoyancy. AIAA Journal. 1966;**4**(10):1751-1758. DOI: 10.2514/3.3773

[14] Zhu T, Cheng R, Lee SA. Continuous-flow ferrohydrodynamic sorting of particles and cells in microfluidic devices. Microfluidics and Nanofluidics. 2012;**13**:645-654. DOI: 10.1007/s10404-012-1004-9

[15] Evrard R, Boutry GA. An absolute micromanometer using diamagnetic levitation. The Journal of Vacuum Science and Technology. 1969;**6**(2):279-288. DOI: 10.1116/1.1492679

[16] Kelland DR, Hiresaki Y, Friedlaender FJ. Diamagnetic particle capture and mineral separation. IEEE Transactions on Magnetics. 1981;**17**(6):2813-2815. DOI: 10.1109/TMAG.1981.1061725

[17] Kendall BRF, Vollero MF, Hinkle LD. Passive levitation of small particles in vacuum: Possible applications to

vacuum gauging. Journal of Vacuum Science and Technology A. 1987;**5**(4): 2458-2462. DOI: 10.1116/1.574871

[18] Mirica KA, Shevkoplyas SS, Phillips ST. Measuring densities of solids and liquids using magnetic levitation: Fundamentals. Journal of the American Chemical Society. 2009;**131**:10049-10058. DOI: 10.1021/ja900920s

[19] Nemiroski A, Soh S, Kwok SW. Tilted magnetic levitation enables measurement of the complete range of densities of materials with low magnetic permeability. Journal of the American Chemical Society. 2016;**138**(4):1252-1257. DOI: 10.1021/jacs.5b10936

[20] Nemiroski A, Kumar AA, Soh S. High-sensitivity measurement of density by magnetic levitation. Analytical Chemistry. 2016;**88**(5): 2666-2674. DOI: 10.1021/acs.analchem.5b03918

[21] Yamamoto Y, Ito A, Kato M. Preparation of artificial skeletal muscle tissues by a magnetic force-based tissue engineering technique. Journal of Bioscience and Bioengineering. 2009; **108**(6):538-543. DOI: 10.1016/j.jbiosc.2009.05.019

[22] Souza GR, Molina JR, Raphael RM. Three-dimensional tissue culture based on magnetic cell levitation. Nature Nanotechnology. 2010;**5**(4):291-296. DOI: 10.1038/nnano.2010.23

[23] Tseng H, Gage JA, Raphael RM. Assembly of a three-dimensional multitype bronchiole coculture model using magnetic levitation. Tissue Engineering—Part C Methods. 2013;**19** (9):665-675. DOI: 10.1089/ten.tec.2012.0157

[24] Tseng H, Balaoing LR, Grigoryan B. A three-dimensional co-culture model of the aortic valve using magnetic levitation. Acta Biomaterialia. 2014;**10** (1):173-182. DOI: 10.1016/j.actbio.2013.09.003

[25] Haisler WL, Timm DM, Gage JA. Three-dimensional cell culturing by magnetic levitation. Nature Protocols. 2013;**8**(10):1940-1949. DOI: 10.1038/nprot.2013.125

[26] Jaganathan H, Gage J, Leonard F. Three-dimensional in vitro co-culture levitation. Scientific Reports. 2015;**4**: 6468. DOI: 10.1038/srep06468

[27] Ilievski F, Mirica KA, Ellerbee K. Templated self-assembly in three dimensions using magnetic levitation. Soft Matter. 2011;**31**(ii):9113-9118. DOI: 10.1039/c1sm05962a

[28] Mirica KA, Phillips ST, Shevkoplyas SS. Using magnetic levitation to distinguish atomic-level differences in chemical composition of polymers, and to monitor chemical reactions on solid supports. Journal of the American Chemical Society. 2008;**130**:17678-17680. DOI: 10.1021/ja8074727

[29] Winkleman A, Perez-Castillejos R, Gudiksen KL. Density-based diamagnetic separation: Devices for detecting binding events and for collecting unlabeled diamagnetic particles in paramagnetic solutions. Analytical Chemistry. 2007;**79**(17): 6542-6550. DOI: 10.1021/ac070500b

[30] Shapiro ND, Mirica KA, Soh S. Measuring binding of protein to gel-bound ligands using magnetic levitation. Journal of the American Chemical Society. 2012;**134**(12):5637-5646. DOI: 10.1021/ja211788e8

[31] Subramaniam AB, Gonidec M, Shapiro ND. Metal-amplified density assays, (MADAs), including a density-linked immunosorbent assay (DeLISA). Lab on a Chip. 2015;**15**(4):1009-1022. DOI: 10.1039/C4LC01161A

[32] Knowlton SM, Sencan I, Aytar Y. Sickle cell detection using a smartphone. Scientific Reports. 2015;**5**:1-11. DOI: 10.1038/srep15022

[33] Tasoglu S, Khoory JA, Tekin HC. Levitational image cytometry with temporal resolution. Advanced Materials. 2015;**27**(26):3901-3908. DOI: 10.1002/adma.201405660

[34] Knowlton S, Yu CH, Jain N. Smart-phone based magnetic levitation for measuring densities. PLoS One. 2015;**10** (8):1-17. DOI: 10.1371/journal.pone.0134400

[35] Baday M, Calamak S, Durmus NG. Integrating cell phone imaging with magnetic levitation (i-LEV) for label-free blood analysis at the point-of-living. Small. 2016;**12**(9):1222-1229. DOI: 10.1002/smll.201501845

[36] Ozefe F, Arslan-Yildiz A. Smartphone-assisted hepatitis C detection assay based on magnetic levitation. 2020 (Unpublished Results)

[37] Daquinag AC, Souza GR, Kolonin MG. Adipose tissue engineering in three-dimensional levitation tissue culture system based on magnetic nanoparticles. Tissue Engineering. Part C, Methods. 2012;**19**(5):336-344. DOI: 10.1089/ten.tec.2012.0198

[38] Tasoglu S, Yu CH, Liaudanskaya V. Magnetic levitational assembly for living material fabrication. Advanced Healthcare Materials. 2015;**4**(10):1469-1476. DOI: 10.1002/adhm.201500092

[39] Türker E, Demirçak N, Arslan-Yildiz A. Scaffold-free three-dimensional cell culturing using magnetic levitation. Biomaterials Science. 2018;**6**(7):1745-1753. DOI: 10.1039/c8bm00122g

Chapter 9

Electromagnetic Levitation of Metal Melts

Vadim Glebovsky

Abstract

The main advantage that attracted the attention of researchers was the lack of contact of liquid metal with refractory lining, which ensured the elimination of one of the main sources of metal contamination by such a harmful impurity, such as oxygen. This is especially important for melting refractory and highly reactive metals and semiconductors. Compared to other melting methods, which also ensured the absence of contact of liquid metal with the crucible (vacuum arc, electron beam floating zone, cold crucible, plasma, etc.), EML of metal melts has a number of significant advantages. Among all types of noncontact technologies, only EML has the functions of levitation and heating.

Keywords: electromagnetic levitation, magnetohydrodynamics, free-surface, cold crucible, liquid metal, numerical simulation, surface tension, oscillations, semi-levitation, inductor, coil

1. Introduction

The creation of new inorganic materials, in particular metallic, requires the improvement of existing and the development of new methods for their preparation. The latest melting methods (vacuum arc, induction and electron beam, electroslag, zone, plasma, laser, etc.) of metal melts contribute to further development of modern materials science, while possessing well-known limitations. There is no universal method of melting that would completely satisfy the most diverse requirements of scientists and engineers engaged in this problem. Therefore, it is necessary to use combinations of known melting methods. In this regard, it is of considerable interest to use new methods of melting, for example, electromagnetic levitation (EML) or induction melting furnaces with a cold crucible. It should be noted that, despite a number of obvious advantages, induction melting with a cold crucible cannot yet be widely applied due to imperfect energy conditions for heating the charge and significant heat losses, although quite a certain progress is observed in the designs of furnaces using the principle of cold crucible.

Otto Muck proposed electromagnetic levitation of a metal melt, which is also called melting in an electromagnetic crucible or noncontact, in 1923. He gave the first and simple theoretical explanation of this phenomenon-the implementation of suspension or metal levitation by an electromagnetic field. However, only 30 years later, the first works on the theory and use of this type of melting appeared. Later, studies appeared aimed at expanding the applied and scientific functions of the levitation of metals and alloys both on a laboratory and semi-industrial scale. It is also interesting that the theoretical justification of the method has always been accompanied by the development of EML [1–11].

- Adjustable residence time of a drop of metal in a liquid state.
- Controlled gas atmosphere and slag phase.
- Controlled metal temperature (from melting temperatures to boiling).
- Ability to use an additional heat source (electronic, laser beam or plasma).
- Vigorous stirring of metal by electromagnetic field.
- Possibility of introducing alloying additives into a liquid drop.
- A favorable ratio between the surface of the droplet and its volume for the passage of heterogeneous reactions.
- Achieving extremely high crystallization rates up to 10^5–10^6°C/s.

Perhaps the disadvantages of the method include a small mass of metal, not exceeding several tens of grams, which to some extent limits the wide industrial application of the method. However, there are known applications of EML in the application of thin coatings in electronics, or the production of fine powders for additive technologies. By the way, for most physical and physicochemical studies, a small mass is not an obstacle.

EML techniques enable the noncontact study of thermophysical properties over a wide temperature range. In this regard, an overview on measuring enthalpy through the method of levitation dropping calorimetry for 30 years is of interest [11]. The results of these measurements of melting enthalpies and melting entropies really amaze with the breadth of coverage (three subgroups of refractory metals) and the complete novelty (obtained for the first time due to EML). For these three subgroups of the periodical table: IVb (Ti, Zr, Hf), Vb (V, Nb, Ta), and VIb (Cr, Mo, W), a group similarity was discovered. Accordingly, the melting entropy of metals of the fourth group is 6.8–8 J mol^{-1} K^{-1}, the fifth group is 10–11 J mol^{-1} K^{-1}, and the sixth group is 13–14 J mol^{-1} K^{-1}, although earlier these data simply did not exist. As for possible industrial applications, the promising process of deposition of thin films of light metals on metal and plastic surfaces, based on the levitation of conductive materials in a high-frequency electromagnetic field, has been studied [12]. The authors focused on the design of the inductor, which ensures the achievement of high specific powers, and the system of rational distribution of vapor, which contributes to the production of a uniform thin coating. According to the authors of [12], this method is most optimal when applying thin coatings of metals with low vapor pressure, such as Al, Ni, Cu and their alloys. It seems quite natural and reasonable to use EML as a part of new technology for high rate physical vapor deposition of coatings onto metallic strip. Many publications are known in which, at a good theoretical level, specific technical problems were solved, for example, increasing the mass of levitated metal or spraying liquid metal [13–22].

During EML, the sample quickly melts, the melt undergoes strong mixing, and as a result, the melt becomes completely homogeneous [23–33]. EML has become widely used to study the refining processes in obtaining metals of high purity, active and with high melting points. EML of metals means the electromagnetic interaction of a sample and a magnetic field. For this, metal samples are placed in an inductor with a high-frequency alternating current, forming an electromagnetic field in which the sample rises and then melts. Induction current, as a rule, arises on the surface of metal samples, and the interaction between the induction current and the high-frequency magnetic field forms the Lorentz force, which is in equilibrium with the gravity of a solid metal sample in a magnetic field of a certain configuration. As a result, the metal sample hangs in the inductor and levitation is realized. The simultaneous action of the induced current or eddy currents generates Joule heat, which leads to heating and melting of the sample.

The lifting power of levitation, temperature and stability are important factors in the state of EML. At different times, the researchers modeled the influence of the design and arrangement of inductors on the dynamic deformation and stability of metal melts, as well as the vibrations of metal droplets at different points in time, for which arbitrary Lagrange-Euler equations and the finite element method were used. The effect of the second (transverse) magnetic field on the stability of rotation of the experimental samples was also investigated. In the process of EML, the samples are simultaneously subjected to heating and levitation, and EML creates a high temperature distributed on and around the levitated melt. The factors affecting the temperature characteristics of EML, the melting of samples with low conductivity and high density at relatively low temperatures are investigated. It seems interesting to study the influence of the structural dimensions of multi-turn inductors and the sizes of a metal sample on the temperature characteristics of EML, analyzed with the aim of correctly choosing a suitable inductor for various applications.

EML of melts is a progressive and universal method for conducting high-temperature physical and physicochemical studies necessary to improve metallurgical processes, as well as a means for producing miniature parts and samples from high-purity metals. Due to its unique characteristics, noncontact levitation provides obvious advantages in the field of research of new materials. Compared to traditional studies using crucibles made of refractory materials, noncontact technology is a unique research technique, and only it opens up the possibility of completely avoiding contaminants entering the metal melt from the refractory material of the crucible. Noncontact levitation is also used to crystallize samples of objects, measure physical and chemical properties, and produce ingots of highly pure crystalline and amorphous materials. Noncontact measurement of the physical and physicochemical properties of liquid metals made it possible to observe thermal and surface vibrations during the levitation of metal melts not only in terrestrial conditions, but also in zero or microgravity [34–38]. Significant underheating of the sample can be one of the advantages of the noncontact method of levitation. Undercooling means the nonequilibrium state of a liquid sample at temperatures below its equilibrium melting point. The noncontact of a liquid sample is the essence of EML, especially when combined with an ultrapure process medium. In addition, EML is one of the oldest noncontact methods of levitation used for material science experiments for decades. New levitation methods include aerodynamic and acoustic levitation, as well as electrostatic levitation. In these methods, levitation forces arise from electrodes located above and below a sample containing a surface charge, while heating is performed, for example, using a laser. EML is the most mature of all noncontact melting methods and has been used for decades in ground-based experiments, as well as in experiments with microgravity with a wide range of alloys.

Thus, surface tension, density and solubility of various gases in liquid metals were measured, as well as decarburization parameters were established, and features of the Fe desulfurization ability of slags in a wide range of carbon concentrations were studied.

2. Physical features of electromagnetic levitation

2.1 Liquid metal floating in an electromagnetic field

A metal with diamagnetic properties can freely hang in a constant magnetic field in the presence of a potential well in it, that is, a region where the tension decreases from the edges to the middle [4–7]. The interaction of a high-frequency magnetic field with a metal leads to the appearance of eddy currents in the latter, which displace the field from the space occupied by the metal, or, in other words, the field inside the metal is weakened by eddy currents. As a result of this, in a variable magnetic field, a nonferromagnetic conductor behaves like a diamagnet in a constant field. Due to the force interaction of eddy currents and the field, the metal is pushed out of the zone with a higher field density to a region with a lower density, that is, into a potential well. If the indicated forces are sufficiently large, then the metal can be raised up despite the action of gravity and held in space in suspension. The simplest field diagram of an inductor system consisting of three parallel wires is shown in **Figure 1**.

It is of interest to consider qualitatively the physical processes that occur during metal levitation. In **Figure 1b**, the location of the metal in the potential well of the magnetic field of three wires with currents is shown. The direction of eddy currents in the surface layer of the metal is opposite to the current in the wires. On the surface of the molten metal, there are special areas characterized by the absence of eddy currents. This is due to the weakening of the magnetic field in these areas. During levitation, in the lower part of the molten metal sample, such areas necessarily exist with almost any configuration of the magnetic field. The absence of eddy currents in such areas eliminates the appearance of electromagnetic pressure, which should lead to the flow of liquid metal through these places, but this does not happen, since the existing surface tension forces compensate for the weakening of the power shell near points 1–4. It should be noted that these forces can balance only a small part of the hydrostatic pressure of the metal and contribute to the retention of the latter at a small height of the liquid column above the indicated surface area. To partially compensate for the weakening of the interaction of eddy currents of the metal with the field, use a two-frequency power supply system of the inductors, which allows you to periodically shift the weakened magnetic field on a larger surface of the metal. Due to the rapid redistribution of the weakened magnetic field and the large inertia of the molten metal, its outflow through singular points does not occur. This technique also contributes to an increase in the mass of the metal. However, the increase in mass is limited by the formation of folds in the lower part of the liquid metal column. The direction of the folds coincides with the magnetic field lines of force, and the depth of the folds is commensurate with the penetration depth Δ.

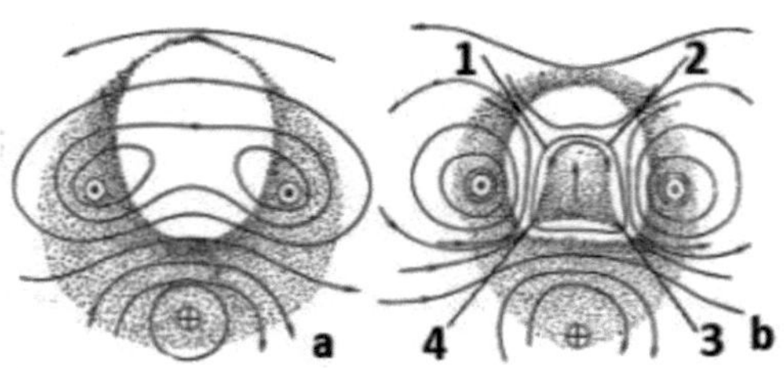

Figure 1.
The magnetic field of three wires with currents (the density of the field lines is shown by the degree of blackness of the shading and fixes the hydrostatic pressure). a—without metal; b—with metal; the numbers 1, 2, 3, and 4 are singular points on the surface of the metal melt.

Along with the indicated limitations of the levitation of metallic melts, there are two more. The first is associated with a given vibrational power in the inductor circuit or, in other words, the limiting field strength corresponds to the limiting metal height h. The second is due to the presence of a minimum volume of liquid metal, which may still be in a levitation state in the cap, close to the penetration depth or skin layer $h_\Delta = \Delta$. This was confirmed experimentally by melting Al, Sn, Fe, Ti and Cu using frequencies of 500–2500 kHz. It should only be noted that the surface tension coefficient α and the penetration depth Δ depend on temperature. Currents flow in the upper wires in the same direction, so the field between them is weaker than around each. A particularly strong field is created near the lower wire, through which the current flows in the opposite direction.

In addition to the above-considered features of the interaction of a high-frequency electromagnetic field and a liquid metal during levitation, there is a group of phenomena associated with the stability of the metal [4–7]. The swaying of a droplet hanging in a magnetic field is not a specific property of the liquid state but is caused by the electromagnetic interaction of the metal with the field. Change in the position of the metal relative to the inductor at a constant emf affects the value of the current flowing through the inductor, which causes a change in the force acting on the metal. When conducting experiments with balls of aluminum floating in air, water and oil, the following features of their behavior were discovered:

- Stable equilibrium with respect to finite disturbances in a more viscous medium than air

- Stable undamped oscillations with a small amplitude of constant magnitude (ball in water)

- Increasing oscillations with an amplitude exceeding the size of the inductor (ball in the air)

- Quickly established stable equilibrium (for all studied balls) in oil

The presented nature of the phenomena does not depend on the current value in the inductor (10–30 A) and on the degree of compression of the balls by the magnetic field. The stability of the metal is ensured if the center of curvature of the surface of the melt in its stable state lies outside the volume of the melt. However, this is impossible, since in acute angles, the value of the Laplace pressure of the curved surface of the liquid would reach infinity. The presence of a special configuration in the electromagnetic field of the potential well, as well as a relatively large volume of metal, leads to the extension of the lower part of the ball, and a drop of metal takes the form of a pear hanging from the cuttings down [7].

A characteristic feature of liquid metal during levitation is intensive mixing inside the drop. A model study carried out with liquid sodium placed in a glass flask, which was in an electromagnetic field (**Figure 2**), showed the existence of turbulent motion of the metal inside the flask [4–7]. Using pitot tubes, as well as photographing methods, melt velocities were measured. It can be seen that the bulk of the liquid in the floating flask moves up. Along the walls of the flask, the metal moves at a much greater speed down. To determine the dependence of the metal velocity on the magnetic field strength, the flask was fixed and the vertical velocity component was measured for various current values. Special experiments without a flask made it possible to conclude that the mixing of the melt during levitation in vacuum or an inert gas is more intense than that described above, since the velocity of the metal on the surface of the drop is not equal to zero. With an increase in the current value in the inductor, the metal velocity increased. The stability of the metal largely depends on the speed of its rotation. This is usually observed with a spherical shape.

The complete process of heating and melting a sample can consist of four stages. At the first stage, the solid sample rises to fixation in a certain stable position (**Figure 3a,b**). This part of the process completely depends on the location of the inductor, field currents and the initial position of the sample. With incorrect levitation parameters, the behavior of the sample may turn out to be unpredictable. The next step is to heat the sample to the melting temperature (**Figure 3b**). At this stage, with increasing temperature, a change in the physical properties of the material itself is possible, which can affect both the change in the electromagnetic field and the stability of the position of the sample. Therefore, this stage is central to the duration and overall effectiveness of the levitation process. The third stage consists of melting the sample (**Figure 3c**). It is known that the melting model is

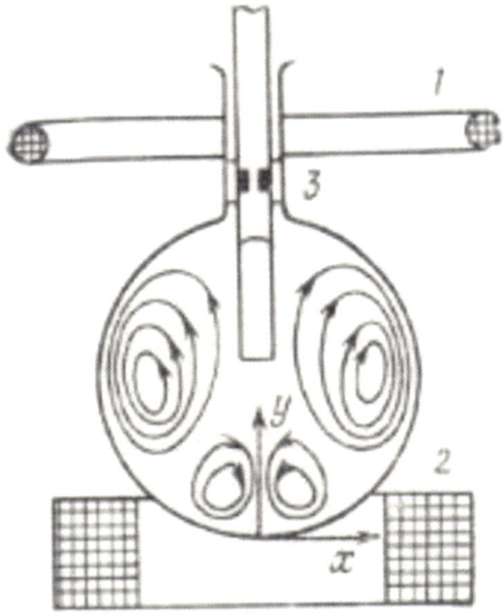

Figure 2.
The motion of the melt in a flask with liquid sodium. 1, 2—inductors with 50 and 700 turns, 3—diaphragm.

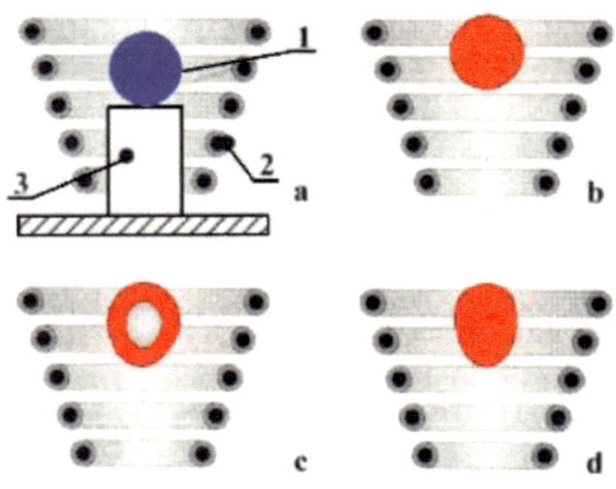

Figure 3.
Simplest arrangement of induction melting with one inductor: 1—solid charge, 2—inductor, 3—ceramic stand. a—the solid sample is fixed in a certain stable position in an inductor; b—the sample is heated to T_m; c—melting the sample; d—the sample is melted, and the melt can be heated to a given temperature due to intensive mixing.

not easy even from a geometric point of view, but it is characterized by the fact that the internal volume of the sample remains solid, while the surface of the sample is covered with a liquid film due to the surface nature of the implementation of Joule losses. At the fourth stage (**Figure 3d**), the sample is melted, and the melt can be heated to a given temperature due to intensive mixing.

2.2 Solid and liquid metal levitation

The main problem of levitation is the development of a theory and its applications to the problem of the retention of liquid metal during EML, although over the past decades there have been many studies that are somehow related to the theoretical basis of levitation. The significant advantages of EML compared to other methods of metal melting have led to the rapid spread of this method; however, this had to be done almost by touch, without sufficient theoretical justification. More recently, more or less adequate theoretical foundations of the method have been developed that can be used to optimize the setup parameters for levitation. Historically, for example, in Russia, the development of levitation occurred along the path of using two-coils and multi-coil inductors. This significantly affected the development of the theory. The main reason for using two types of inductors is the different power of high-frequency generators used for levitation. Two-coil inductors require increased power (26–100 kW), while multi-coil inductors are able to operate at lower power (8–15 kW). Theoretical prerequisites and experimental possibilities for using two-coil inductors (with parallel turns) were developed by Alex Vogel and his lab [6, 39]. In accordance with the development of this lab, the effect of an idealized uniform electromagnetic field on the metal half-space is summed up from the electromagnetic force (F) and the power absorbed by the metal (P_s), which goes to heat it. Analytically, this is expressed as the following relationship:

$$F = \frac{1}{2}\sqrt{\mu / \pi \rho f}\, Ps\ \mathrm{S} \tag{1}$$

where F is the electromagnetic force acting on the metal and equal to its mass; μ is the magnetic permeability of the vacuum; ρ is the electrical resistivity of the metal; f is the field frequency; and Ps is the power transmitted to the metal and referred to a surface unit:

$$Ps = \frac{1}{2}\sqrt{\pi \mu \rho f}\, H^2 \tag{2}$$

where H is the amplitude of the magnetic field on the surface of the half-space. As a result of solving the system of Maxwell equations for a plate in a longitudinal plane-parallel magnetic field, the dependence of the magnetic component of the field on its frequency is established for

$$\beta x = Re\left[\frac{\delta y Bz}{2}\right] = \mathrm{Const.} \tag{3}$$

and a constant plate thickness. This dependence was hyperbolic in nature. Therefore, with a fixed size of the metal sample, there is a well-defined frequency range at which the metal theoretically levitates in an electromagnetic field. The choice of a specific frequency value is determined by the required sample temperature. In addition, it is also necessary to take into account the configuration of the field in order to determine the nature of the dependence of the lifting force on its parameters. Several similarity criteria were theoretically established (conditions for the equality of electromagnetic pressure and mass, equality at specific points of hydrostatic and Laplace forces, and equality relating the skin effect, circular frequency, etc.) and experiments were conducted to simulate the vaporization of liquid aluminum on molten sodium. In an elongated two-coil inductor "with a parallel reverse coil," a group of droplets that did not merge with each other due to the existence of repulsive forces between them was stably held, which is obviously explained by the interaction of currents of the opposite direction flowing at the ends of the droplets. For levitation, when the field is inhomogeneous and the metal moves in it, falling into zones of various configurations, coefficient A characterizing the configuration of the field is introduced into Eq. (2), and then Eq. (2) takes the form:

$$F = \frac{1}{2}\sqrt{\mu / \pi \rho f}\; A P s \tag{4}$$

The validity of this equation is verified experimentally, provided that the inductor has a coefficient value A = Const. The power supplied to the copper, molybdenum and niobium balls with a diameter of 15 mm at different field frequencies using the same two-coil inductor "with side parallel coils" was calorimetrically determined. This power was compared with the calculated one [6, 39]. The difference between P_s obtained by calculation and experimentally was 23%, which proves the good reliability of Eq. (4), in which the values of coefficient A are determined experimentally for each type of inductor. For a two-coil inductor "with two side parallel coils," the values of A vary from 0.7 to 0.9, whereas for a two-coil inductor "with two parallel coils," these values vary from 0.4 to 0.7. For a two-coil inductor "with sequential switching of coils," the values of coefficient A are in the range of 0.2–0.7. Ensuring the transmission of a given power P_s depends on the mass of the metal, the surface of the sample, the frequency of the field and a certain coefficient A. Coefficient A characterizes the degree of heterogeneity of the electromagnetic field. It was experimentally established that the greater its heterogeneity, the smaller the coefficient A, which theoretically can tend to zero, but it is practically impossible to get it less than 0.2. The heterogeneity at the surface of the melt is different; therefore, the value of A also depends on the position of the melt in the inductor and on the size and shape of the sample. When considering the behavior of a metal ball in the field of a two-coil inductor system, the functional dependence of coefficient A on the position of the sample was determined [6]. All investigated metals are divided into three groups in accordance with the interval of fixed temperature:

Group 1: Al, Fe, Co, Ni, Cu, Rh, Hf, Ir ($t_r > t_m$); Group 2: Ti, Zr, Nb, Mo, Ru ($t_r > t_m$, depending on I and U); Group 3: Ta, W, Re, Os ($t_r < t_m$).

For a two-coil inductor, the dependence of r/A on the volume of various metals is obtained, which varies similarly to the previously described [6]. The increased temperature range corresponds to the optimal volume of the metal, so that depending on the frequency, a different temperature of the metal melt is set, and in vacuum, the temperature is always higher. It is promising to obtain a stable melt temperature using two fields: holding and heating, but not for all metals equally. For Group 1 metals, the use of two-frequency heating is excluded, since the confining field overheats these metals much higher than t_m. For metals of Groups 2 and 3, two-frequency heating is practically possible.

The interdependence between the values of the function F characterizing the skin effect and the power P transmitted to the metal is extremely important. Obtaining a given metal temperature is achieved either by changing the frequency of the generator or by choosing the shape of the inductor. A change in the current in the inductor cannot lead to a direct change in the P/F ratio, since both quantities depend on I^2. With increasing current, the metal in the inductor rises and falls into the area with a lower electromagnetic field strength and a large gradient of it. This means that temperature can only be controlled to a limited extent.

In addition to the dependences of the lifting force described above on the frequency and intensity of the electromagnetic field, as well as on its configuration, there are specific conditions for a limited choice of the field frequency for holding a liquid metal, which is characterized by a change in shape. It was previously noted that metal during levitation is located in a potential well in which at least one singular point or area of a weakened field necessarily exists through which the metal does not pour out only due to surface tension on a curved surface. In areas of a weakened field, the pressure of a liquid metal column is balanced by the difference in surface tension values on the curved surfaces of the lower and upper parts of the metal [6]. The use of a two-coil inductor at the selected frequency and configuration of the electromagnetic field revealed the need to study the relationship between the voltage applied to the inductor and the behavior of the liquid metal during levitation. Experimentally, the lowering of a drop was recorded with a decrease in voltage. When a certain value was reached, the metal began to flow out of the inductor. The release of the melt can be controlled by increasing the area of the weakened field while maintaining the strength of the magnetic component of the field, necessary to hold the bulk of the metal. It was experimentally established that in a two-coil inductor there are three zones of the electromagnetic field [6]. In the first zone, the solid metal hangs, and the liquid merges independent of the capillary constant. In the second zone, the liquid metal hangs unstably; its degree of stability depends on the volume of the metal and does not depend on the capillary constant. In the third zone, the position of the metal in the inductor is associated with the presence of volume dependence. The use of a multi-coil inductor at the selected frequency and configuration of the electromagnetic field revealed the need to study the relationship between the voltage applied to the inductor and the behavior of the liquid metal during levitation.

2.3 Temperature of the levitated melt

Along with the retained metal melt, the production and regulation of its temperature are of great importance. For two-coil inductors, the theoretical foundations and technological designs that provide the necessary heating of the samples were considered in [6, 7]. The validity of the functional dependence (4), which relates the electromagnetic lifting force to the power absorbed by the metal, has been proved experimentally. In the steady state in vacuum, the power transmitted to the metal is equal to the radiated power. For most metals, the temperature dependence of the power P_s radiated from a unit surface is well known (**Figure 4**). This power is usually determined by the reverse calculation. It is known that ensuring the transmission of a given power P_s depends on a number of factors: the mass of the melt, the surface of the sample, the frequency of the field and coefficient A. The mass of the melt during levitation in a multi-coil inductor can be determined by knowing the minimum frequency of the electromagnetic field that implements the levitation of a metal melt with a given height and physical properties.

Obtaining high temperatures during the levitation of metal melts has no fundamental obstacles. In practice, this is accomplished by choosing, for example, a multi-coil inductor, the conical part of which is open at a small angle and whose field has a small tension gradient. To obtain the necessary lifting force, a sufficiently large current is needed, due to which the metal is heated. Due to the fact that the power increases proportionally with frequency, and the lifting force is much less dependent on it, the field frequency is increased to obtain a higher temperature. In each case, the field frequency must be chosen so that the value $X = R/\Delta$ (here R is the radius of the sphere and Δ is the thickness of the skin layer) is more than 10 and the value of F does not depend on the change in ρ. Otherwise, during heating, the

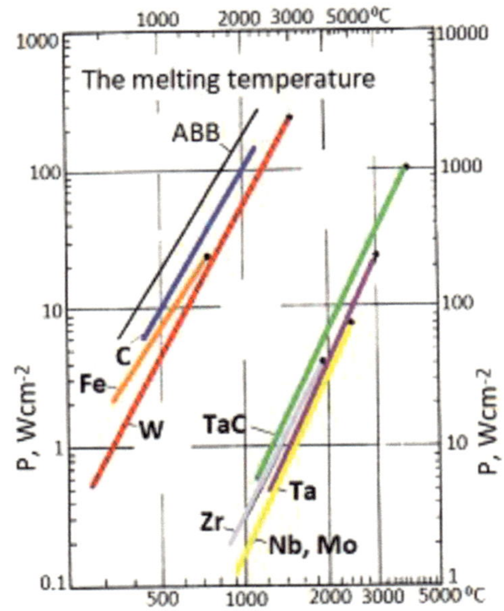

Figure 4.
Dependence of the specific radiation power on the temperature of refractory metals. ABB—the absolutely black body.

value of X may become less than 10, and the value of F at I = Const. will decrease, which will lead to the discharge of the molten drop. Low temperatures of the melt can be obtained at such a frequency and diameter of the spherical sample, when the value of X is more than 10. In this regard, F should be about 50% less than its full value (when X is more than 10). As mentioned above, the current in this case is regulated within certain limits, and the shape of the inductor must be such that the intensity gradient is maximum. In this case, the stabilizing forces become very small.

Temperature control can be done in two ways. The first is the selection of such an inductor shape in which there is a strong dependence of the P/F and F ratios on the position of the sample inside the inductor. An inductor is suitable for this, in which there exists a possibly large linear decrease in field strength along the axis of the inductor. In this case, P also depends linearly on the position of the sample. If the current is increased, the sample rises to the upper part of the inductor and its temperature decreases. The second method consists in the separate action of two inductors fed by currents of two frequencies: the lower one-for levitation of the metal at the lowest temperature-and the upper one-only for heating and melting the sample with regulation of current and frequency. In EML, samples simultaneously participate in heating and soaring, so that the power of EML unambiguously indicates a high temperature distributed around the soaring sample. This explains the interest in studying factors regarding the influence on the temperature characteristics of EML with the final goal of gaining additional knowledge in order to not only fully realize the levitation and melting of samples with low conductivity and high density at relatively low temperatures, but also fully use the primary role of inductors on the temperature characteristics of the electromagnetic field. So, in [6, 39] for the first time, using the finite element analysis, the influence of the design of inductors and the mass of samples on the temperature characteristics of levitation were studied. The modeling capabilities are confirmed experimentally as a result of a detailed analysis of the effect of the inductor on the temperature characteristics of levitation when choosing the optimal inductor in various applications. The shape of the metal melt depends on the degree of compression of the melt by an electromagnetic field, which is largely determined by the configuration of the inductor. In a multi-coil inductor, the limiting melt height can be approximately two times smaller than in a two-coil inductor [6].

To obtain a given stable steady-state temperature, not only two-frequency levitation, cooling gas mixtures, but also other heating sources (electron beam, light beam, electromagnetic beam with a frequency corresponding to the centimeter and millimeter wavelength ranges) can be used. Refractory metals are known (Group 3), which, under experimental conditions, levitated in an inductor, but were not melted. Therefore, for their additional heating, a source of thermal energy-an electron beam-was used [6, 7]. Indirect heating is also used due to the relatively low efficiency of induction heating, which is mainly due to the presence of a large gap between the inductor and the metal sample and, in fact, depends on the design of the inductor. The use of additional heating is shown by the example of tungsten, which should melt at a frequency of 440 kHz, a power of 160 kW and a mass of 28 g. **Figure 5** shows experimentally determined temperatures depending on their mass and voltage at the inductor (40–90 V). Obviously, an increase in the mass of the sample cannot lead to a significant increase in the temperature of the metal (~3000°C). At this temperature, the power emitted by the melt surface (~7.5 cm^2, the shape of the melt is a spinning top) is 1.7 kW, which corresponds to 1.7% of the power consumed by the generator. This complicates the cooling of the inductor with water and significantly increases the voltage (up to 300 V). The arrow in **Figure 5** shows the temperature rise of a metal after being heated by an electron beam (~3700°C). Radiated power increases to 3.4 kW, that is, 1.7 kW is additionally transmitted, which is 68% of the total power of the electron beam unit (2.5 kW).

Similar results were obtained with EML of a 30 g Nb sample in a multi-coil inductor at a voltage of 30 V, a generator power of 25 kW and a frequency of 80 kHz. The metal temperature was 2100°C, which in terms of absorbed power is 0.5 kW or 2% of the power consumed by the generator. The metal was heated by an electron beam to a melting point of ~2415°C, and the power was 1.2 kW. The melt surface emitted ~1 kW. This means that the electron beam additionally transmitted at least 0.5 kW or 40% of the power consumed by the electron beam setup. By the way, electromagnetic levitation of the Nb melt without additional heating can be carried out in the same type of inductor using a more powerful generator (60 kW) with a frequency of 440 kHz and a voltage of 160 V.

Additional heating may also be necessary in specific cases, for example, with silicon levitation. Pure silicon is known to have an extremely high electrical resistance; therefore, at ordinary frequencies (of the order of hundreds of kilohertz) and power (tens of kilowatts), silicon cannot be levitated. With increasing temperature, the electrical resistance of silicon decreases, especially sharply at the melting temperature; therefore, for silicon levitation, it is necessary to pre-heat it with an electron beam. If the field frequency during silicon levitation is about 200 kHz, then the sample should be heated to 10,000°C, and at frequencies of 70–80 kHz, up to 14,000°C.

Two-frequency heating has been studied in less detail. The experiments with aluminum were carried out in air. An inductor for EML was connected to a machine generator with a frequency of 8 kHz, and an inductor for heating, from a generator with a frequency of 440 kHz. Five different two-coil inductors were tested, producing a two-frequency field. The most rational were (1) an inductor with a single coil for holding the sample, placed above the main heating coil; (2) an inductor with a single holding coil placed between two main heating coils; and (3) an inductor with a single holding coil placed above the main heating coil.

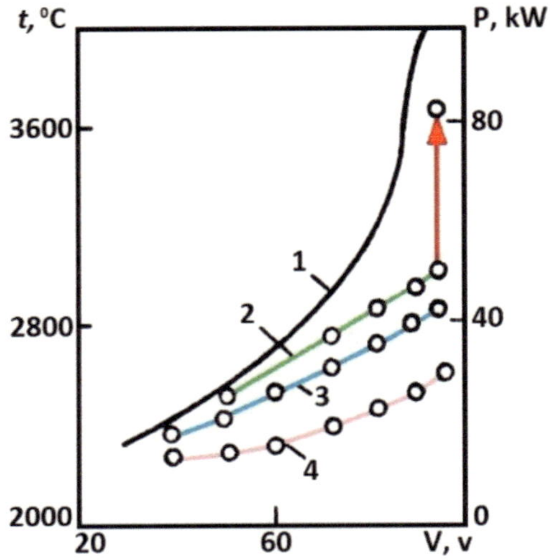

Figure 5.
The dependence of the power of the generator and the temperature of tungsten of various masses on the voltage at the inductor; the arrow shows the temperature of the metal after additional heating by an electron beam: 1—P, kW; 2–28.5 g; 3–20 g; 4–10 g.

In axisymmetric EML, the Lorentz force vanishes on the axis of symmetry. At the lowest point on the axis of the levitated sample, only the surface tension of the melt prevents the flow of the melt, which affected the limitation of the mass of the melt-not more than 50–100 g. With electromagnetic two-frequency levitation, lifting forces also arise along the axis of the levitated sample. Digital models developed to optimize EML were implemented in [13–18]. At the same time, data were obtained on the electromagnetic flux and the dynamics of the free surface, for which calculations of the electromagnetic force, as well as modeling of the melt volume and restoration of the shape of the free surface, were made.

2.4 Setups for EML of melts

Depending on the voltage at the inductor, its design and the conditions of the technology and experiment, all setups for EML are divided into two groups: with an internal inductor and an external inductor relative to the reaction vessel. The first group includes setups powered by quenching circuit generators and equipped with two-coil or multi-coil inductors of all known types when operating in a vacuum or inert gas environment. The second group includes setups with multi-coil inductors, powered by generators without a quenching circuit for operation at atmospheric pressure or low discharge. The setup with a two-coil inductor, designed to obtain samples used in metal research, appeared in the middle of the last century [6, 39]. The metal vessel housed an inductor for levitation, a rotary table with copper molds, a table for initial samples and a manipulator. Later, the same authors [6, 39] developed a 27-position setup for levitation, consisting of a high-frequency generator, a reaction vessel and a vacuum unit. Another reaction vessel is shown in **Figure 6b**. A sample with a manipulator was placed in a double-coil inductor, where levitation and melting of the metal took place. A transparent shutter 6 protected the sight glass 8 from condensation of metal vapors on it. After a given exposure, the melt was poured into the mold 2, located coaxially with the inductor 4. A characteristic feature of such setups was the presence of a rotating table with molds and a manipulator.

Another feature of such setups was devices for various physicochemical studies, for example, the interaction of elements in the melt-slag-gas system. The reaction vessel shown in **Figure 6** is made of copper, and the flanges of the vessel are closed with plexiglass covers. Water cools the inductor and the stabilizer ring, located directly above the conical multi-coil inductor. A copper pin crystallizer cooled by liquid nitrogen is intended for crystallization of the drop with a liquid slag (in the lower part of the drop). The mold could be moved in vertical and horizontal planes without violating the tightness of the reaction vessel. The setup was powered by a 10 kW generator with power regulation at the inductor by lowering the primary coil, which provided finer regulation.

Typical for setups of the second group with an external inductor relative to the reaction vessel is the use of multi-coil inductors, quartz glass for the reaction vessel and various means for instant crystallization of metal melts (**Figure 7a,b**). This is explained by the fact that such plants were used for high-temperature studies of the solubility of gases in melts, in particular, nitrogen in iron-carbon melts and the pressure of saturated iron vapor. The body of the reaction vessel is made of quartz glass with polished joints of individual elements. The location of the reaction vessel in the inductor is characteristic-its configuration repeats the internal shape of the inductor, which led to a slight increase in the diameters of the upper and lower coils of the inductor and, as a result, to a decrease in the efficiency of the inductor and an increase in the used power of the generator. At the bottom of the reaction vessel was a turntable with molds for melt crystallization. The turntable was rotated in such a way that there was a free socket on the same axis as the reaction vessel, which made it possible to measure the temperature of the melt using an optical pyrometer. Depending on the purpose of the experiment, some elements were changed in the setups of the second group and additional devices were introduced, for example, to crystallize the melt when oxygen was falling in the atmosphere or to crystallize at an increased velocity using the hammer-anvil device.

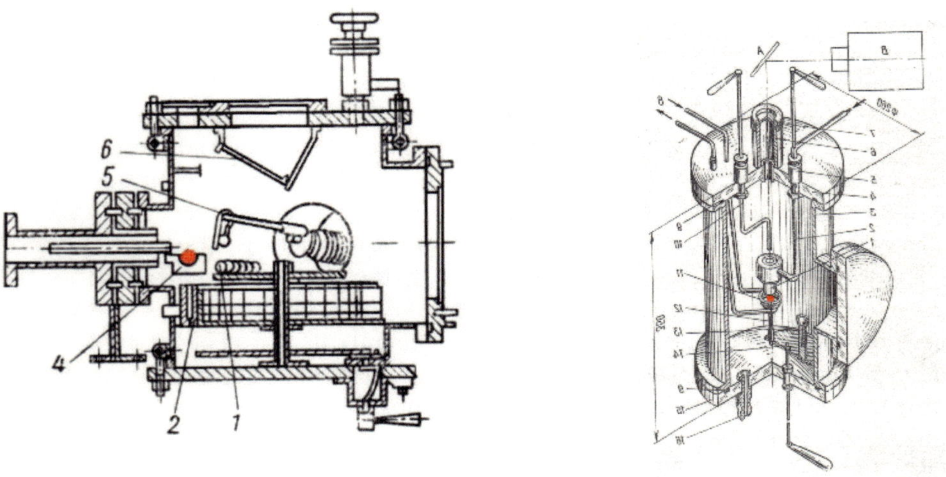

a b

Figure 6.
a—Multi-position setup for levitation and studies in systems "melt-gas"; b—Levitation setup for physicochemical studies in "melt-gas" and "melt-slag-gas" systems.

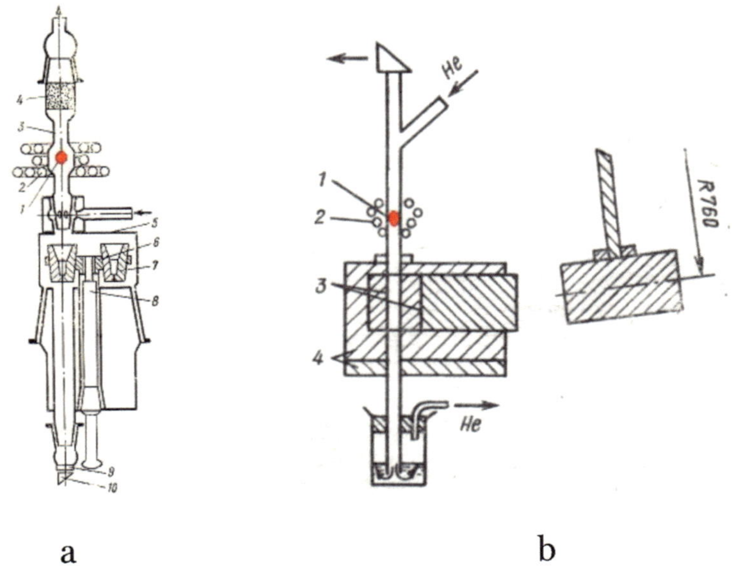

Figure 7.
Schemes of levitation quartz setups for studying the solubility of gases in liquid metals with fixation of dissolved gas using: a—Crystallization in molds; b—crystallization with a hammer and anvil.

2.5 Inductor designs

The inductor and the heated metal sample form a single electromagnetic system, similar to a transformer in short circuit mode. However, the functions of its parts are clearly separated in the transformer: electric current passes through the windings, magnetic flux through the magnetic circuit. In contrast, the surface layer of the heated sample is simultaneously a secondary electrical winding and part of the magnetic circuit. Therefore, in the general case, when calculating the parameters of the inductor, it is necessary to take into account not only the magnetic flux passing in the gap, but also the flux in the metal. In addition, the consideration is also complicated by the fact that the values of ρ and μ at different points of the cross section of the heated metal are different and change over time. The heating process (cold, intermediate and hot modes) during electromagnetic levitation and the assumptions made to simplify the relationship between ρ and μ for subsequent calculations ($\rho \times \mu$ = Const.) are studied in detail and are available in the reference literature. The values of μ are determined as a function of the magnetic field strength at the interface, using the magnetization curve. Due to the fact that the magnetization of the magnetic field depends on the specific power in the heated sample, the magnetic permeability is its function. Examples of the general calculation of the inductor are quite possible, and the necessary relations can be obtained from solving the electromagnetic field equation as applied to the propagation of electromagnetic energy inside a flat conductor of infinite thickness. Examples of calculating a single-coil quenching cylindrical inductor, with which the diameter of the inductor and its width, voltage and current, the efficiency of the inductor and the power supplied to it could be determined. However, it is impossible to use this calculation to determine the parameters of inductors for EML, since it does not take into account the main difference between EML and known heating methods-the existence of a force supporting a metal sample in solid and liquid states.

In this regard, the most important and necessary feature of EML is the use of special inductors, the electromagnetic field of which holds and heats the metal sample. As noted above, the great merit of Alex Vogel [6] and his lab consists in the development of two-coil inductors, consisting of two parallel coils connected in parallel. The inductor design is shown in **Figure 8**. The vertical bends of the profiled copper tube are made for stable melt levitation. One of the important operating parameters of two-coil inductors is the relationship between the power referred to the mass of the metal and the square of the amplitude of the magnetic field.

In addition, a feature of this inductor is the presence of two critical voltages: the first and second, indicating a limitation of stability in the lower and upper positions. It also has some disadvantages: manufacturing difficulties (profiled copper tube); the maximum achievable temperature of the melt is always lower than in the inductors of two other designs; the need to place an inductor inside the reaction vessel. However, this inductor has several advantages: its dimensions are relatively small; the field is symmetrical; potential difference is minimal; the bottom of the inductor is at the same potential; and the mass of the sample is greatest. Two-coil inductors of two other types differ significantly in their characteristics from the inductor described above. Their designs are presented in **Figures 9** and **10**. In both cases, the P/G values monotonically increase with increasing power supplied to the inductor. A higher temperature of the metal melt is achieved as a result of a larger compression of the melt by the field. However, this leads to an increase in hydrostatic pressure in the melt and levitation of a smaller volume of metal compared to the first inductor.

The inductor of the second type has the following characteristic features: (1) the maximum potential difference between the inputs is lower than in the third inductor; (2) the bottom of the inductor is not equipotential, although the low voltage at the lower coil and the large contact resistance between the inductor and the sample exclude melt welding; (3) higher losses in current leads than in the third inductor; and (4) the limiting temperature of melt during levitation is always less than in the third inductor. For the inductor of the third type, the characteristic of the inductor of the second type is valid with the difference that: (1) the potential difference between the inputs of the inductor is maximum; (2) losses in current leads are minimal; (3) the highest sample temperature; (4) the smallest sample mass [7].

Multi-coil inductors having a reverse coil or a stabilizing ring are most widely used in the practice of electromagnetic levitation. **Figure 11** shows the multi-coil inductor with a stabilizer ring, which was used for several physical studies with Nb, Mo, Fe, Co and Ni [7]. The main advantages of these inductors are as follows: (1) the possibility of using generators of low (8–15 kW) and medium (30–60 kW) power; (2) obtaining an electromagnetic field of almost any configuration; (3) a small

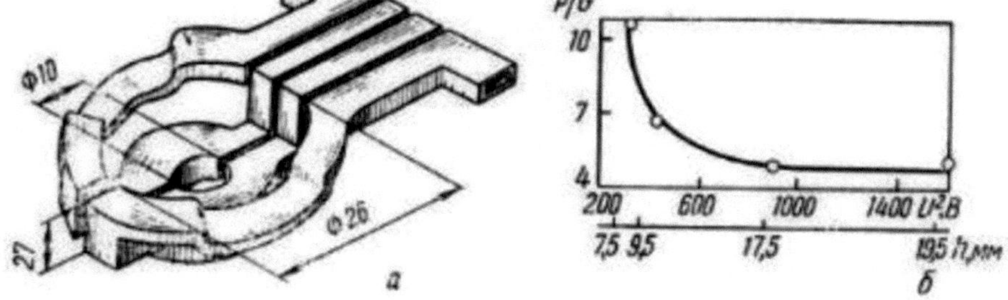

Figure 8.
Type 1 (a) inductor and the dependence of the power transmitted to the sample on the voltage across the inductor (b); h *is the distance from the center of the sample to the upper plane of the lower part of the coil, and the field frequency is 200 kH.*

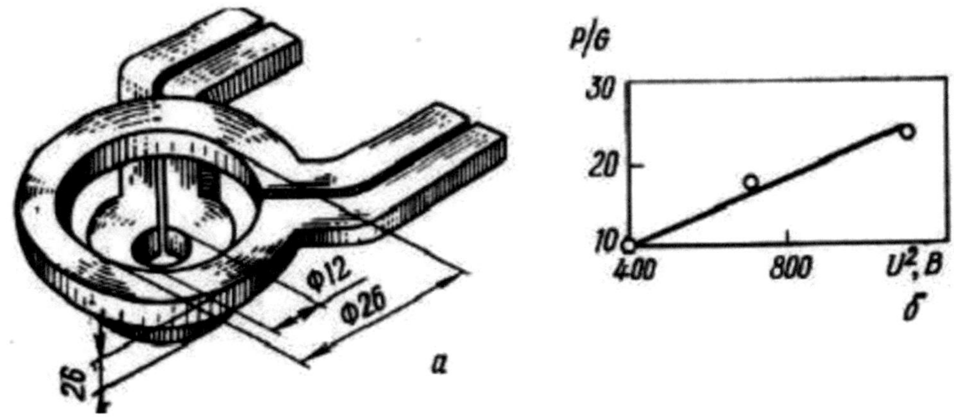

Figure 9.
Type 2 (a) inductor and the dependence of the power transmitted to the sample on the voltage across the inductor (b); the distance from the center of the sample to the plane of the lower coil of the inductor is constant, and the frequency is 200 kHz.

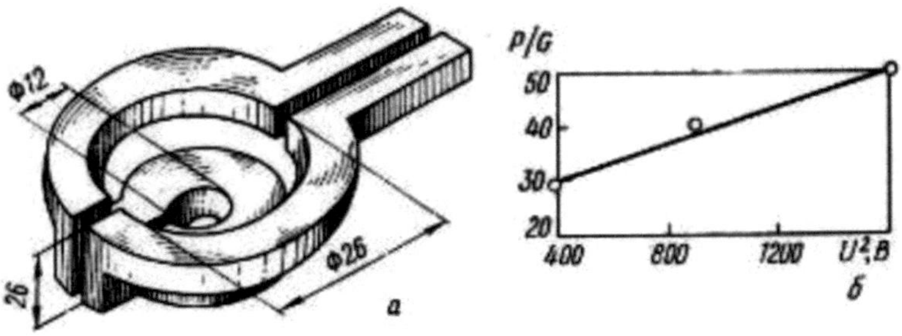

Figure 10.
Type 3 (a) inductor and the dependence of the power transmitted to the sample on the voltage across the inductor (b); the distance from the center of the sample to the plane of the lower coil of the inductor is constant, and the frequency is 200 kHz.

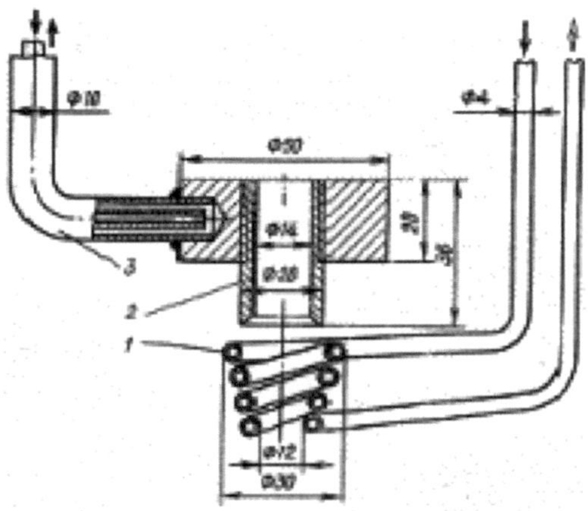

Figure 11.
Multi-coil inductor with stabilizing water-cooled ring. 1—Inductor, 2—stabilizing ring, 3—coaxial water cooling.

potential difference at the inductor when using generators with a quenching circuit; (4) the accuracy with which inductors are made is not limited; (5) relative ease of manufacture, no shaped tubes, special welding; and (6) ease of operation. The disadvantages are also simple: (1) low mass of the sample; (2) limited temperature control; and (3) the possibility of burning the inductor with the melt.

Previously, the behavior of the melt during levitation in a multi-coil inductor (the number of coils of the conical part, the angle of inclination and the number of reverse coils) and the parameters of the levitated sample (levitation and melt temperature) were examined in detail. A rigorous calculation of a multi-coil inductor for levitation of a melt is still impossible. In this regard, the selection and manufacture of such inductors are carried out empirically, taking into account general ideas about their work. The literature describes more than two dozen designs of multi-coil inductors used for various physical studies. It is almost impossible to classify them according to any criteria, since the tasks solved by their creators were always different. The technique for selecting multi-coil inductors is as follows [7]: the selection criteria are stable levitation (determining the location of the sample in the potential well of the electromagnetic field) and the maximum possible regulation of the temperature of the metal (e.g., solid copper and liquid iron). A generator with a power of 30 kW and a frequency of 230 kHz was used. Inductors were prepared from a copper tube with an outer diameter of 4 and 6 mm. The initial shape, the number of coils of the conical part and the cone angle (~60°) were selected in accordance with the data of [7]. The internal diameters of the lower coil of the conical inductor and the upper return coil (d_1 = 15.5 mm; d_2 = 22 mm) were the same for all multi-coil inductors and were selected in connection with the required metal mass (average weight 3.5 g) and dimensions of the reaction vessel. Stable levitation of the melt was determined visually: the time from the moment of melting to the outflow of the metal from the inductor was noted (the latter phenomenon is associated with the saturation of liquid copper with oxygen, which reduces the surface tension of the copper melt during levitation in air). It turned out that along with the size d_1, the number of spiral coils affects the lifting force during levitation.

3. EML in physical research

The emergence of new metallurgical processes, such as electroslag melting, electron beam melting, arc melting, induction vacuum melting and plasma melting, revealed the limitations of the available thermodynamic and kinetic data necessary for the correct refining of liquid metal. A characteristic feature of these methods are higher temperatures in comparison with the temperatures at which traditional methods of melting steel and alloys are carried out. The melting temperature of high alloyed steels can reach 1700–1750°C, and temperatures up to 2000–2500°C develop in the reaction zone when the steel is purged with oxygen or air. When melting refractory metals in the arc and electron beam setups, local overheating of the metal is possible at 1000°C above the melting temperature. There are almost no experimental data on the behavior of liquid metals at such high temperatures, which is explained by the limited capabilities of experimental methods. Electromagnetic levitation significantly extends the temperature range of studies related to the solubility of gases in liquid metals, the processes of interaction of metal and slag melts with the participation of the gas phase, etc. It is known that such studies are impossible due to chemical reactions developing between the melt and the refractory material of the crucible.

3.1 Physical properties and chemical reactions studied by EML

The study of heterogeneous systems with the help of EML encounters a number of difficulties with which it was not necessary to deal with the study of liquid metals by traditional methods. One of the main difficulties is the strong evaporation of the metal. This process arises due to two characteristic features of the electromagnetic levitation method: high temperature and a large specific surface of a molten metal drop (**Figure 12**). Evaporation is particularly active when the liquid metal is held in a deep vacuum. For example, the temperature of the carbon iron melt can reach 2000–2100°C in a few seconds, and its five-minute exposure in vacuum is 10^{-5}–10^{-6} mm Hg and leads to a decrease in mass by 3–4 times. Such intense evaporation of metal in a vacuum makes it difficult to study liquid metals by electromagnetic levitation with prolonged exposure.

To reduce the contribution of evaporation processes during electromagnetic levitation, an inert gas or gas mixtures are used at different pressures, which can be changed from a few mm Hg to several atmospheres. Although metal evaporation proceeds more slowly in this case, metal vapors condense not only on cold surfaces, but also in less heated areas of the reaction vessel, forming condensate flakes. Their appearance makes it difficult to observe the liquid drop and distorts the temperature measurements. In addition, chemical reactions can occur between the gas phase and the metal vapor, which distort the results. Since the temperatures of the liquid metal droplet and the gas phase are different, two processes affect the rate of metal evaporation: natural gas convection and vapor condensation. Both of these processes, together or separately, increase the rate of evaporation with an increase in the temperature gradient in the system. Around the drop of molten metal, a boundary gas layer appears, the thickness of which can vary from 0.06 cm at an ambient temperature of 0 K to 0.32 cm at an ambient temperature equal to the temperature of the melt. If the surface of a molten metal, for example, iron, has a temperature of ~2000 K, then at the boundary of the gas layer, it will be 1650 K at an ambient temperature of 0 K. In this regard, when studying heterogeneous processes using electromagnetic levitation, it is necessary to take into account the possibility of reactions between metal vapors and the gas phase in the boundary gas layer at temperatures different from the parameters of the molten metal.

Intensive evaporation of the metal of a liquid droplet leads to strong dusting of the sight glasses. In those cases when the experiment is conducted in a gas atmosphere, the evaporating metal does not pollute them; however, the condensate formed in the chamber, along with the gas flows rising from the liquid droplet, continuously moves along the reaction chamber. For this reason, the temperature of a liquid droplet is usually measured from the side or from the bottom through special devices installed in the reaction vessel. For example, to measure the temperature of liquid iron under conditions of strong evaporation, you can use a

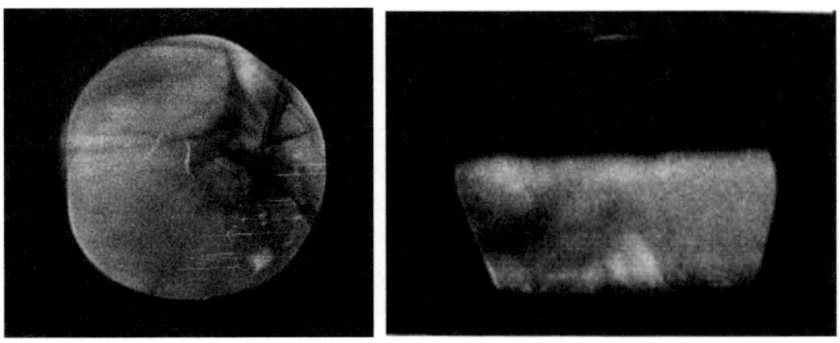

Figure 12.
Top and side views of the same Ø 10 mm liquid iron drop in the multi-coil inductor. The top view is almost a sphere with a disfigured surface, more like a dried apple, but during the shooting on the surface of the sphere, the author observed traces of that "muscle play" under the covers that took place in the volume of the sphere. The side view also contains no evidence of the ideal surface of a metal drop during levitation [7].

movable tube mounted on bellows and brought almost to the melting inductor on the side. When measuring temperature at the bottom of a melt drop, the pyrometer is sighted through an opening in the turntable. After performing several operations to measure the temperature, a mold is placed under the drop, which is installed in another nest of the table.

The microdistribution of temperature over the sample was studied, and the measurement was carried out in the central and peripheral zones of a solid sample heated to premelting temperatures. The difference in temperature measurements of these sections did not exceed 10–12°. Apparently, in the case of molten metal, the temperature difference becomes completely insignificant due to intensive mixing and continuous updating of the surface of the liquid droplet. This eliminates the accumulation of impurities on the surface of the droplet, which could affect the temperature measurement.

The temperature control of liquid metal during levitation can be carried out by optical, radiation and color pyrometers. The use of the first two is allowed in the absence of noticeable smoke or contamination of the sight glasses. The surface of the metal when measuring temperature should be free of oxide or any other film. Color pyrometers are not very sensitive to the appearance of films on glasses; therefore, they are used especially often, although they are difficult to operate. In the case of using a brightness pyrometer, it is necessary to first measure the emissivity of metals in a special setup, since a strong dependence of the emissivity on temperature has been revealed. To determine the temperature dependence of the emissivity of some refractory metals in the solid state, it is desirable to use a model of an absolutely black body. The shape of such samples should be chosen so that the samples had the greatest stability in the inductor and the direction of the hole in an absolutely black body looked vertically upward. The calibration of the pyrometers was carried out on the reference points of pure metals, molten during levitation or in a small ceramic crucible. A drop of metal was slowly melted and then crystallized. To do this, you can take advantage of a change in the gas flow rate or a copper crystallizer cooled by liquid nitrogen, supplied to the drop from above. When it comes into contact with metal, it is possible to observe the crystallization front and make multiple measurements of the melting point. When using pyrometers with automatic recording of the melting and crystallization temperatures, they are fixed with the corresponding areas or bursts (**Figure 13**) [7]. The use of conventional thermocouples is difficult, since it is difficult to eliminate the influence of a high-frequency electromagnetic field on the junction and, secondly, the volume of the metal is usually not large enough to neglect the errors introduced in the temperature readings when a solid junction is introduced into a liquid metal drop.

3.2 Measurement of surface tension and melt viscosity during levitation in zero gravity in parabolic flights and ISS

Volumetric metal glasses or amorphous alloys are a new phenomenon in materials science. The main advantage of these materials is their superior mechanical properties compared to conventional crystalline materials. Amorphous alloys are solid metal materials with a disordered glass-like structure of an atomic scale. They are formed when their cooling from a liquid state occurs much faster than the critical cooling rate. During supercooling of the melt, the increasing thermodynamic driving force of crystallization and amorphous atomic kinetics compete with each other. A strong increase in viscosity during cooling and a high probability of amorphization of the melt also establish boundary conditions for the correct choice of parameters of the process. To justify superplasticity, it is important to know the temperature-dependent viscosity of the alloy. Thus, in order to create technology

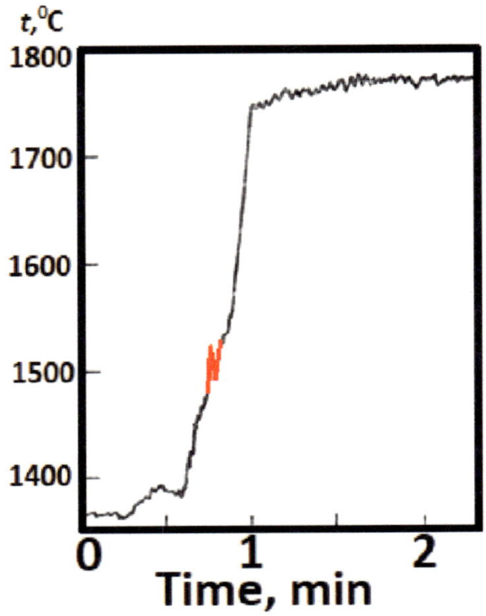

Figure 13.
Temperature of liquid iron in the process of electromagnetic levitation; red—a "zig-zag" at the moment of melting the iron sample.

and theoretical models, it is necessary to know the basic thermophysical properties in a wide temperature range. Electromagnetic levitation is clearly a powerful technique for the noncontact manipulation of electrically conductive samples. This method allows to correctly measure the surface tension and viscosity of metal melts. However, under conditions of gravity of the earth, the melt in natural geometry or raised by an electromagnetic field will be significantly deformed. The simultaneous control of temperature and levitation is limited under normal gravitational conditions of 1 G, since the electromagnetic field needed to lift the samples can heat the sample to significant temperatures, even above the melting point. Flows in a heated, deformed melt drop under terrestrial conditions are poorly controlled (laminar transition to turbulent), which makes it necessary to conduct experiments in zero gravity (microgravity) conditions. One of the possibilities to achieve microgravity in a short period of time (10–20 s) is parabolic flights, for example, performed using the Airbus A310 or International Space Station. The experimental results were obtained during several parabolic flight campaigns in 2016 and 2017 using the TEMPUS EML setup. The surface tension and viscosity of the Pd43Cu27Ni10P20 melt were measured, and it was shown that the temperature dependence of the viscosity of this alloy can be well described by the free volume model. In addition, in [23–33], using X-ray diffraction, it was confirmed the absence of the long-range order characteristic of amorphous samples. The morphology of the samples was analyzed using X-ray computed tomography before and after flights.

The noncontact of a liquid sample is the essence of electromagnetic levitation, combined with an ultra-clean environment. In addition, electromagnetic levitation is one of the oldest noncontact methods of levitation used for materials science experiments for decades. Electromagnetic levitation is the most mature of all noncontact melting methods and has been used for decades in ground-based experiments, as well as in microgravity experiments with a wide range of alloys. Electromagnetic levitation in terrestrial conditions has some problems associated with gravitational forces, so if levitation is performed under microgravity conditions, then only small levitation forces are required to compensate for residual accelerations. With an inductor optimized for microgravity, heating and positioning of the samples can be carried out almost independently. Since positioning requires minimal effort, the low temperature mode is more accessible, convection is reduced and the deformation of the samples is eliminated. As a result, many additional samples can be processed, and their thermophysical properties can be determined with higher accuracy.

Other factors are the length of periods of microgravity and the ability of the experimenter to interactively control the experiment during a parabolic flight. Electromagnetic levitators have been developed for a wide range of carriers and tasks. Due to the flight duration, which was on the order of 1–2 weeks, it was possible to conduct several experiments lasting several hours each, where each experiment with this sample included several melting cycles with this sample. The experiments were preprogrammed, so interactive access via telemetry is possible. The ISS missions are essentially a continuation of previous Spacelab missions and provide long-term access to a good microgravity environment. Several batches of samples can be loaded onto an object, processed and returned to the ground. The experiments are preprogrammed; interactive access via telemetry is possible.

3.3 Atomization of liquid metals in levitation

The advent of 3D metal printing and other additive technologies has stimulated an increase in demand for spherical metal powders with high rheological flow characteristics. The spraying process consists of feeding a vertical sacrificial rod into a conical inductor, where the end of the rod is melted by eddy currents of an electromagnetic field, resulting in the formation of a stream or droplets of liquid metal that are sprayed with a powerful flow of inert gas. In fact, one of the functions of classical electromagnetic levitation is involved in the process-the melting of the metal, without holding it by a magnetic field. The spraying unit is very simple and consists of a feeder with a sacrificial rod, a melting chamber with an inductor, a spraying chamber with nozzles for supplying an inert gas, a powder storage device and a generator. The proposed method is noncontact and ideal for producing high-purity, reactive and refractory metal powders. All process parameters are known and easily adjusted, which allow full control of the size of the powders. The process is simple, manageable and flexible. Perhaps, this process stands out among analogues for its simplicity and reliability, especially in the production of high-quality pure spherical powders from refractory and rare metals such as titanium, zirconium, niobium and precious metals, which are in great demand in additive technologies in aerospace, medical and other industries [20–22].

3.4 Chemical equilibrium in the system metal-slag-gas during EML

Heating, melting and crystallizing a metal melt with slag occur in a controlled gas atmosphere or in vacuum. The exposure time to the onset of chemical equilibrium in the "iron melt-slag melt-gas" system is usually short and does not exceed several minutes. Slag melting occurs due to levitation and heating of iron, liquid slag initially covers a metal drop with a thin film, which can be observed visually, and then it collects in the lower part of the drop and is held in liquid state by interfacial tension. It is this joint behavior of the molten metal and slag plus convection in liquid iron that ensures the rapid achievement of chemical equilibrium in the distribution of sulfur (the usual or radiochemical sulfur isotope ^{35}S). The initial sample for levitation was a capsule of a specially prepared alloy of iron with carbon weighing

~5 g with an opening in which a slag powder weighing 0.35 g was laid; the hole was tightly closed with a thin foil lid of the same alloy. The initial sample was placed in an inductor for subsequent levitation, which, depending on the task, was carried out in vacuum, inert gases and carbon monoxide. After the generator was turned on, levitation and melting of the sample took place with heating to the experimental temperature, and the metal melt took an egg-shaped form, on the lower half of which slag was collected. At the beginning of the experiment, when the capsule was melted, the slag was in the form of a thin film enveloping the entire drop. The experimental time required to reach chemical equilibrium was determined by diffusion in the liquid slag phase.

Using levitation, the sulfur distribution between iron-carbon melts and lime-alumina slag was studied [7, 28, 32]: the dependence of the equilibrium sulfur distribution coefficient on the C content, gas phase composition and temperature.

Experimental results are shown in **Figure 14**. Comparison of the reduced coefficients of the equilibrium distribution of sulfur, obtained by numerical modeling using known thermodynamic data and the experimental equilibrium values of the reduced coefficient of the sulfur distribution, is shown in **Figure 15**. Since the known data for the equilibrium constant of the desulfurization reaction differ by one and a half orders of magnitude, this noticeably affects the reduced sulfur distribution coefficient. On the whole, it is necessary to recognize such a discrepancy as completely admissible and justified. As can be seen in **Figure 15**, in the logarithmic coordinates, the calculated and experimental curves slightly differ in slope, which is due to tolerances in the calculation and experiment [7].

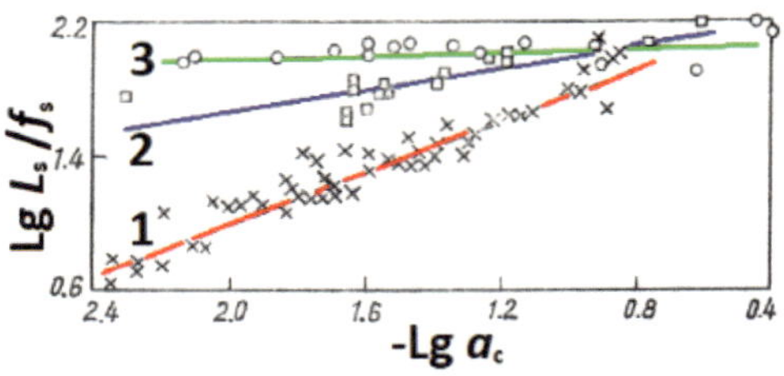

Figure 14.
Experimental dependence of the reduced coefficient of distribution of ^{35}S between Fe-C and oxide slag melts on the activity of C in iron. 1—P_{Co} = 1 atm, 2000°C; 2—P_{He} = 1 atm, 1750°C; 3—P_{Ar} = 1 atm, 2000°C.

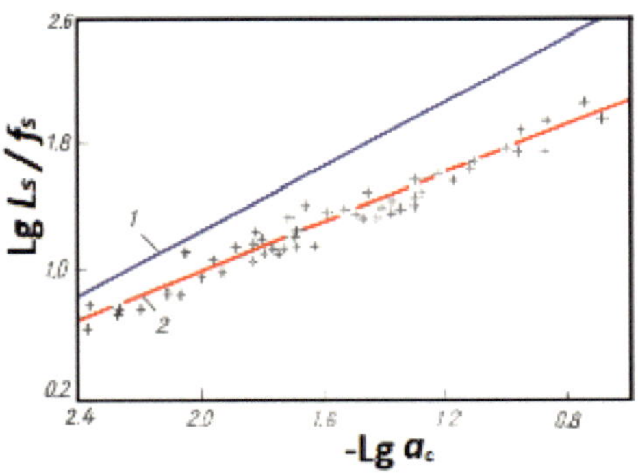

Figure 15.
Dependence of the experimental (2) and calculated (1) coefficients of the distribution of sulfur between liquid Fe-C and slag on the activity of C in iron at P_{co} = 1 atm and 2000°C.

3.5 Reaction of C + O → CO in the melts of Fe and Nb at EML

3.5.1 Melts of Fe-C-O at EML

For comparing EML with crucible melting, the study was made of the influence of the refractory lining on the decarburization kinetics of Fe-C-O samples, which were carried out on melts in corundum crucibles, degassed in vacuum at 1700°C. The data obtained showed that the composition of the products of the decarburization reaction is close to equilibrium; however, the total volume of gases released from the metal depends on the melting method. In melts in a crucible, the gas evolution of CO and CO_2 always exceeded the gas evolution during levitation of similar samples. For stable levitation of samples of this system, a setup was used with an inductor inside the chamber; the working pressure in which was 10^{-7} atm and a metal temperature of 2000 ± 30°C. The received dependence (see **Figure 16**, area 6) is qualitatively confirmed by the data obtained from the smelting of Fe-C-O alloys in an electron beam setup at 1550°C and a vacuum of 10^{-7} atm (see **Figure 16**, area 3). The calculated dependence of the O content on C for 2000°C confirms the disproportionate relationship between the deoxidation capacity of carbon and the partial pressure of CO in the gas phase. The observed decrease in the O concentration in the metal is due to a change in the deoxidizing ability of C due to a change in P_{CO}. Apparently, the data obtained for melting in an inert atmosphere characterize the maximum possible increase in the deoxidizing ability of C at 2000°C.

The increase in the deoxidizing ability of C dissolved in liquid iron, which occurs as a result of a decrease in P_{co} over the melt, is observed to a certain limit determined by the kinetics of the CO bubble growth. Since the deoxidizing ability of C is usually expressed by the product of the concentrations of C and O in the metal m = [% C] × [% O], the dependence of the deoxidizing ability on the partial pressure of CO can be expressed by a parabola. Under levitation conditions, when liquid metal does not come into contact with a refractory lining and there is no influence of the lining on the formation of CO bubbles, volume decarburization should prevail, not excluding the evaporation of CO molecules from the surface of a liquid droplet. It is likely that the contribution of evaporation to the removal of CO from a levitated drop of liquid metal should decrease as O adsorption in the surface layer decreases. In low-carbon iron, the formation of CO bubbles in a metal volume is facilitated by thermodynamic and kinetic factors. If the nucleus is comparable in size to large nonmetallic inclusions, the value $2\delta/r$ is much lower than atmospheric pressure. In this regard, as the partial pressure of CO in the gas phase decreases, the

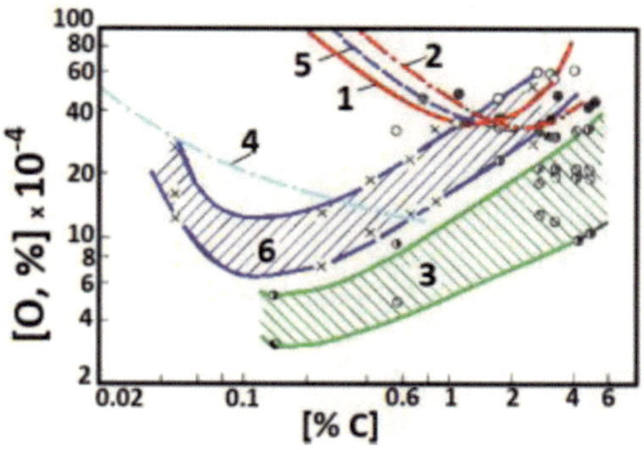

Figure 16.
Solubility O in Fe-C-O melts. 1—P_{co} = 1 atm, 2000°C; 2—P_{co} = 1 atm, 2000°C; 3—P_{co} = 10^{-7} atm, 1550°C, EBM; 4—P_{co} = 10^{-7} atm, 1600°C (refractory crucible); 5—P_{co} = 1 atm, 1600°C (refractory crucible); 6—P_{co} = 10^{-7} atm, 2100°C.

decarburization reaction intensifies until this pressure is equal to the partial pressure of CO in the gas nucleus and the deoxidizing ability of C becomes constant. The surface decarburization of carbon iron cannot be decisive due to the low diffusion rates of C and O to the metal-gas interface. When bubbles form in the metal volume, the pressure necessary to overcome the forces of surface tension should be two to three orders of magnitude higher than atmospheric. It follows that in iron-carbon melts, in particular at contents >1.5% C, with a decrease in the partial pressure of CO in the gas phase, the deoxidation ability of C in iron should not change. The description of EML experiments on Co and Ni is not given here because of limited space in the chapter.

3.5.2 Melts of Nb-C-O at EML

The interaction of C and O should be taken into account when obtaining pure metals and studying their surface and bulk properties. However, the study of the behavior of these impurities in solid and liquid metals was often qualitative because of the complex nature of physical and chemical processes in the bulk and on the surface of the metal and because of significant experimental difficulties. A model was created within the framework of which a closed system of equations was obtained [34, 35]. The model allows describing the kinetics of the interaction of C and O during maintaining them in a liquid state in vacuum. It was found that if C and O impurities with initial concentrations of $N_O(0)$ and $N_c(0)$ are uniformly dissolved in the metal volume, then by thermal desorption of CO and MO molecules in vacuum (at high contents of oxygen, MO_2), since the average O concentration $N_O(\mathrm{t})$ decreases infinitely, then, in accordance with the proposed model, the average C concentration $N_c(\mathrm{t})$ should reach a definite threshold level $N_c(\infty)$. It is shown that the ratio between the average concentrations of O and C is uniquely fixed by two parameters, N^* and S. The first of them, $N^* = \omega/G$, is determined by the ratio of the constant of the MO desorption rate ω to the effective constant of the CO desorption rate G and has a certain critical value, which sets the characteristic concentration scale. The effect of one impurity on another becomes high only when its concentration beats this critical value. The parameter S is a dimensionless indicator of the relative intensity of diffusion or surface processes.

The relative simplicity of the model makes it possible to experimentally check the behavior of the average concentrations of carbon $N_c(t)$ and oxygen $N_O(t)$ in time, as far as the created model [34, 35] corresponds to reality and determines the parameters N and S. So, if the model is correct, then in the coordinates $\partial\Delta/\partial$ (In N_c) $- \Delta$, regardless of the ratio of the initial concentrations of $N_c(0)$ and $N_o(0)$, the experimental points should be on a common line with the fixed slope and ordinate.

The comparison of the results with the theoretical ones becomes much easier if the S values in the experiment are sufficiently large, which means that diffusion is more intensive than surface processes. In this case, the dependence In $N_c = f(\Delta)$ should be close to linear (**Figure 17**). Thus, the experimental points for different ratios of the initial concentrations O and C in the In $N_c - \Delta$ coordinates should fit on parallel straight lines with a slope $1/N$.

The kinetics of the interaction of C and O was investigated in a wide range of ratios of their initial concentrations during levitation of Nb and Mo in vacuum. The temporal relationship between the average concentrations of C and O in a wide range Δ follows a simple law. The kinetics of conducting of the average concentrations of C and O was calculated numerically, which showed a good alignment between the digital simulation and the experiment for all studied series. The effective constant of the CO desorption rate and the high-temperature sticking

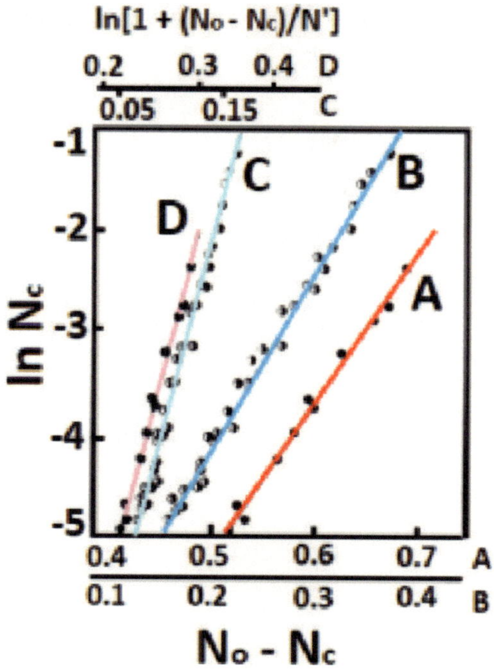

Figure 17.
Mean C concentrations as a function of the value $\Delta = N_o - N_c + N_c^{(\infty)}$*, plotted in coordinates In* $N_c - \Delta$ *(lines A and D) and In* $N_c - In(1 + \Delta/N')$ *(lines C and B) on the basis experimental data.*

coefficient of CO to Nb and Mo were received. The description of EML experiments on Mo is not given here because of limited space in the chapter.

4. Conclusions

The main advantage that attracted the attention of researchers was the lack of contact of liquid metal with refractory lining, which ensured the elimination of one of the main sources of metal contamination by such a harmful impurity, such as oxygen. This is especially important for melting refractory highly reactive metals and semiconductors. Compared to other melting methods, which also ensured the absence of contact of liquid metal with the crucible, EML of liquid metals has a number of significant advantages: adjustable residence time of a drop of metal in a liquid state; controlled gas atmosphere and slag phase; controlled metal temperature (from melting temperatures to boiling); ability to use an additional heat source (electron beam, laser beam or plasma); vigorous stirring of metal by electromagnetic field; possibility of introducing alloying additives into a liquid drop; a favorable ratio between the surface of the droplet and its volume for the passage of heterogeneous reactions; and achieving extremely high crystallization rates up to 10^5–10^{6}°C/s.

The noncontact of a liquid sample is the essence of EML, combined with an ultra-clean environment, which is an excellent instrument for researches. In addition, ELM is one of the oldest noncontact methods of levitation used in materials science experiments for decades. EML is the most mature of all noncontact melting methods and has been used for decades in ground-based experiments, as well as in microgravity experiments with a wide range of alloys. EML in gravitational conditions has some problems associated with gravitational forces, so if levitation is performed under microgravity conditions, then only small levitation forces are required to compensate for residual accelerations. With an inductor optimized for microgravity, heating and positioning of the samples can be carried out almost independently.

Conflict of interest

The author declares no conflict of interest.

Author details

Vadim Glebovsky
Institute of Solid State Physics RAS, Chernogolovka, Russia

*Address all correspondence to: glebovs@issp.ac.ru

References

[1] Okress EC, Wroughton DM, Comenetz G, Brace PH, Kelly JCR. Electromagnetic levitation of solid and molten metals. Journal of Applied Physics. 1952;**23**:545-552

[2] Lohöfer G. Theory of an electromagnetically levitated metal sphere. SIAM Journal of Applied Mathematics. 1989;**49**:567-581

[3] Lewis JC, Neumayer HRJ, Ward RG. The stabilization of liquid metal during levitation melting. Review of Scientific Instruments. 1962;**39**:569-575

[4] Vuzens AP. The use of physical modeling to study the soaring of liquid metal over parallel coils. Magnetic Hydrodynamics. 1971;**3**:19-22

[5] Kirko IM. Liquid Metal in the Electromagnetic Field. Moscow: Energy Publishing; 1964. p. 160

[6] Vogel AA, Pavlov NA, Korkin IV, Sidorova TA. Inductors for heating and melting metals by levitation. Russian Metallurgy (Metalli). 1961;**5**:51-61

[7] Glebovsky VG, Burtsev VT. Levitation Melting of Metals and Alloys. Moscow: Metallurgy Publishing; 1974, p. 176

[8] Zeitsch K. Rotary electromagnetic levitation for high-temperature testing. High Temperatures-High Pressures. 1970;**2**:419-428

[9] Nordine PC, Weber JKR, Abadie JG. Properties of high-temperature melts using levitation. Pure Applied Chemistry. 2000;**72**:2127-2136

[10] Busse FH. Oscillations of a rotating liquid drop. Journal of Fluid Mechanics. 1984;**142**:1-8

[11] Frohberg MG. Thirty years of levitation melting calorimetry—A balance. Thermochimica Acta. 1999;**337**:7-17

[12] Baptiste L, van Landschoot N, Gleijm G, et al. Electromagnetic levitation: A new technology for high rate physical vapor deposition of coatings onto metallic strip. Surface & Coatings Technology. 2007;**202**:1189-1193. DOI: 10.1016/j.surfcoat.2007.05.075

[13] Zhu XR, Harding RA, Campbell J. Calculation of the free surface shape in the electromagnetic processing of liquid metals. Applied Mathematical Modelling. 1997;**21**:207-214

[14] Lv Q, Jiang H, et al. Simulation analysis and experimental study of the temperature characteristics of electromagnetic levitation. AIP Advances. 2018;**8**:105127

[15] Spitans S, Jakovics A, Baake E, Nacke B. Development, verification and application of numerical model for coupled free surface and liquid metal flow calculation in EM field. In: 7th International Conference on Computational and Experimental Methods in Multiphase and Complex Flow: MPF 2013: 3-5 July, 2013. A Coruna, Spain: Elsevier B.V. WIT Transactions on Engineering Sciences. Vol. 79. 2013. pp. 49-61

[16] Spitans S et al. Numerical modelling of free surface dynamics of melt in an alternate electromagnetic field. Part II: Conventional electromagnetic levitation. Metallurgical and Materials Transactions B. 2015;**47**:522-536

[17] Spitans S et al. Numerical modelling of free surface dynamics of melt in an alternate electromagnetic field. Part I: Implementation and verification of model. Metallurgical and Materials Transactions B. 2013;**44**:593-605

[18] Spitans S et al. New technology for large scale electromagnetic levitation melting of metals. Magnetohydrodynamics. 2015;**51**:121-132

[19] Spitans S. Investigation of turbulent free surface flow of liquid metal in electromagnetic field [PhD thesis]. Riga: University of Latvia; 2015

[20] Spitans S, Franz H, Baake E. Numerical modelling and optimization of the electrode induction melting for inert gas atomization (EIGA). In: Proceedings of 11th PAMIR International Conference. "Fundamental and Applied MHD", (Reims, France). 2019. pp. 327-331

[21] Okomura T, Shibata T, Okochi N. Production of gas atomized Ti alloy powder by levitation melting furnace with electromagnetic nozzle. In: Ninomi M et al., editors. Ti-2007 Science and Technology. Tokyo, Japan: The Japan Institute of Metals; 2007. pp. 997-1000

[22] Okumura T, Yamamoto K, Shibata M. Large scale cold crucible levitation melting furnace with a bottom tapping nozzle. In: Proceedings of 6th International Conference "Electromagnetic Processing Materials" (Dresden, Germany). 2009. pp. 521-524

[23] Egry I, Diefenbach A, Dreier W, Piller J. Containerless processing in space—Thermophysical property measurements using electromagnetic levitation. International Journal of Thermophysics. 2001;**22**:569-578

[24] Fecht HJ, Wunderlich RK. Fundamentals of liquid processing in low earth orbit: From thermophysical properties to microstructure formation in metallic alloys. Journal of the Minerals Metals & Materials Society (JOM). 2017;**69**:1261-1268

[25] Hofmann DC, Roberts SN. Microgravity metal processing: From undercooled liquids to bulk metallic glasses. NPJ Microgravity. 2015;**1**:15003

[26] Lohöfer G, Piller J. The new ISS electromagnetic levitation facility: MSL-EML. In: Proceedings of 40th AIAA Aerospace Science Meeting. 2002. DOI: 10.2514/6.2002764

[27] Higuchi K, Fecht HJ, Wunderlich R. Surface tension and viscosity of the Ni based superalloy CMSX-4 measured by the oscillating drop method in parabolic flight experiments. Advanced Engineering Materials. 2007;**9**:349-354

[28] Easter S, Bojarevics V, Pericleous K. Numerical modelling of liquid droplet dynamics in microgravity. Journal of Physics: Conference Series. 2011;**327**:012027

[29] Wunderlich RK, Fecht HJ, Lohöfer G. Surface tension and viscosity of the Ni based superalloys LEK94 and CMSX-10 measured by the oscillating drop method on board a parabolic flight. Metallurgical and Materials Transactions B. 2017;**48**:237-242

[30] Egry I, Lohöfer G, Jacobs G. Surface tension of liquid metals: Results from measurements on ground and in space. Physical Review Letters. 1995;**75**:4043-4046

[31] Zhao G, Chen S, et al. Single-phase electrorheological effect in microgravity. Soft Matter. 2011;7:7198-7200

[32] Seidel A, Soellner W, Stenzel C. EML—An electromagnetic levitator for the international space station. Journal of Physics: Conference Series. 2011;**327**:012057. DOI: 10.1088/1742-6596/327/1/012057

[33] Mohr M, Wunderlich RK, et al. Surface tension and viscosity of liquid Pd43Cu27Ni10P20 measured in a levitation device under microgravity. NPJ Microgravity. 2019;**5**(1):4-12. DOI: 101038/s41526-019-0065-4

[34] Glebovsky VG, Shipilevsky BM. Kinetics of oxygen and carbon removal from liquid molybdenum in the process of high-frequency levitation in vacuum. Journal of Alloys and Compounds. 1992;**184**:297-304

[35] Glebovsky VG, Kapchenko IV. Shipilevsky BM kinetics of interaction of carbon and oxygen in liquid niobium during levitation melting. Journal of Alloys and Compounds. 1992;**184**:305-313

[36] Burtsev VT, Glebovsky VG, et al. The effect of carbon and the gas phase on the distribution of sulfur between liquid iron and lime-alumina slag. Russian Metallurgy (Metalli). 1968;**6**:43-50

[37] Burtsev VT, Glebovsky VG. A new method to study the distribution of impurities between metal and slag during non-contact electromagnetic levitation. Industrial Laboratory. 1965;**31**:1352-1356

[38] Burtsev VT, Glebovsky VG, et al. The study of the deoxidizing ability of carbon in molten iron, cobalt and nickel by levitation at 1550-2000°C and $P_{CO} < 1$ at. In: Ageev N, editor. Kinetics and Thermodynamics of Interaction of Gases with Liquid Metals. Moscow: Nauka Publishing; 1974. pp. 142-145

[39] Smirnov VV, Korkin IV. Levitation of electrically conductive samples of large volumes and masses. Russian Metallurgy (Metalli). 1978;**1**:89-95